Nonscience Returns

NONSCIENCE RETURNS

WITH THE PSEUDOTRANSMOGRIFICATIONALIFIC
EGOCENTRIFIED REORIENTATIONAL
PROCLIVITIES INHERENTLY INTRACORPORATED
IN EXPERTISTICAL CEREBROINTELLECTUALISED
REDEPLOYMENTATION WITH SPECIAL REFERENCE
TO QUASI-NOTIONAL FASHIONISTIC
NORMATIVITY, THE INDOCTRINATIONALISTIC
METHODOLOGICAL MODALITIES AND SCALAR
SOCIO-ECONOMIC PROMULGATIONARY
IMPROVEMENTALISATIONALISM PREDELINEATED
POSITOTAXICALLY TOWARD
INDIVIDUALISTIFIED MASS ACCEPTANCE
GRATIFICATIONALISTIC
SECURIPERMANENTALISATIONARY
PROFESSIONALISM,
or HOW TO RULE THE WORLD

BRIAN J. FORD

Curtis Press

© Brian J. Ford 2020

ISBN (hbk) 978-0-9934002-5-4
ISBN (ebook) 978-0-9934002-6-1

Original edition published in 1971 by Wolfe Publishing, London, UK
Revised edition published in 2020 by Curtis Press, Great Yarmouth, UK

Cover design: JudithSDesigns&Creativity, *www.judithsdesign.com*

Typeset by Falcon Oast Graphic Art Ltd, *www.falcon.uk.com*

Distributed in North America by SCB
Distributed in the United Kingdom and the rest of the world by Gazelle Book Services Ltd

Visit Curtis Press at *www.curtis-press.com*

Contents

Dedicated,
with love,
to Jan
(and also to Alfie, for all his inspiration)

'Where once lay tangled wastes of sickly green,
Can now proud works of Nonscience clear be seen'
ANON.

Looking to the time—only a few years hence—when the whole world's surface can be developed and properly regulated, the strident and stimulating manifestations of Applied Nonscience stand proud above the overgrown, weedy wastes of yesteryear.

Publisher's note

The original *Nonscience* dates from 1971, and caused a sensation. It was translated, featured on television, and enthusiastically reviewed. To celebrate its fiftieth birthday it is being republished, with updates for each chapter to show how its predictions came true—and the COVID-19 pandemic makes it particularly timely.

This extraordinary book reveals a world dominated by Experts. For these all-powerful people, public image and media exposure are all that matters. Scientists, devoted to discovering the truth, have been superseded by Experts who use confusing language to dominate us and lay claim to colossal grants in their quest for power. Integrity and objectivity are gone; opportunism and duplicity reign.

With the internet, there's no need for schools—they've become a state-funded baby-sitting service for working parents. Why do youngsters go to university? Not to broaden their minds, but to stay up all night, get drunk, and get laid. Going to uni is the most painless way of leaving home, and teenagers then borrow huge sums of money to fund their university. A university chief can earn five times as much as the Prime Minister. Experts rule the banks and, when the system collapsed, government bailed them out with borrowed money so that they could pay themselves huge bonuses, as before. Professor Ford had a much better idea what to do with the money—which would have provided a stimulus to the economy.

Experts study weird things, like a bird called *Bugeranus*, a fungus *Spongiforma squarepantsii,* a beetle called *Agra cadabra*, and *Pieza rhea*, a fly. They are all real! There are articles like 'Fifty Ways to Love Your Lever' and 'Fantastic Yeasts and where to find Them', and papers with multiple authors (in 2015 *Nature* published one with 5,154 authors). Encyclopaedias copy facts from each other, and are dotted with mistakes, so you will find biographies of Dag Henrik Esrum-Hellerup and Lillian Virginia Mountweazel—invented to fill the pages. Neither was real. Plagiarism is rife. Even reputable organisations like the Royal Society and Cambridge University now steal published ideas, trying to claim them as their own. In some countries, one-third of research has been copied from somebody else.

Experts prey on the public who are ignorant of what's going on and they ensure that we are surrounded by fake news. The Amazon is not the 'lungs of the world' (it contributes no oxygen whatever to our atmosphere) and our hysteria about plastic is similarly misplaced. The BBC transmits bogus science programmes because it is acceptable to boast that you 'can't understand maths', or 'don't know physics', though nobody would admit 'I never heard of Shakespeare' or 'I'm ignorant of art'. When computerised planes crash or ships ram the dockside because they are controlled by computers, it is the crew who get blamed, though the real culprits are the youngsters who wrote the computer code (we never hear about them).

Experts say they use long words to aid communication, but Ford reveals that the terms are really there to keep outsiders at bay. Experts take decisions that kill people, yet are immune to blame—saying 'lessons have been learned' means they're off the hook. British people say they don't want American chicken, and wouldn't eat chlorine-washed food. Yet they do, every day. They approve of quiche, while avoiding a fried breakfast—even though the ingredients are similar, and the quiche can be more deadly. People follow those bake-off programmes, though the fatty food they promote kills people. Ford believes these shows should have a health warning and is surprised we don't have the 'Great Tobacco Smoking Challenge' or the 'Blindfold Railway-Crossing Elimination Game'.

This book should be read by everybody wishing to understand the modern world. Huge enterprises (like the Human Genome Project and the Large Hadron Collider) have conned us out of billions of pounds, while smaller teams had better results at a fraction of the cost. It is time to call a halt to this global confidence trick—and *Nonscience Returns* is the book that will guide us.

THE AUTHOR

Brian J. Ford has frittered away his life creating a new multidisciplinary approach to science. He is the person who first introduced laws governing the safe handling of viruses and bacteria, and went on to demonstrate intelligence in microbes. He has answered puzzles like the evolution of giant dinosaurs, spontaneous human combustion, coagulation of the blood, the development of the microscope, and scores of other topics. He edits encyclopaedias, was science editor for *Guinness World Records* (and wrote their *New Quiz Book*); presented his own BBC shows Science Now and Where Are You Taking Us?, hosted Food for Thought on Channel Four, and even had his own television game-show; he has visited most countries in the world and has lectured internationally for decades—the leading cruise lines have him as their celebrity speaker. Professor Ford has appeared on Today and Any Questions? while writing reports on the EU's nuclear research, bed-bugs, conservation, algae, locust breeding, and Lithuanian politics. He appeared on the satirical programme Week In, Week Out, has done stand-up comedy, written for *New Scientist* and *Scientific American*, and writes a controversial column in *The Microscope* journal (search for ford CF01.htm). Brian J. Ford has connections with many universities and the popular An Evening With Brian at the Inter/ Micro conference in Chicago has run for over 30 years. His books (approaching 40 of them) have been published in about 150 editions around the world. Curtis Press approached him about a new book, but all he'd agree to was a reprint of one he'd written earlier. Typical.

Introduction

You're a member of the public? Do not read this book—put it back at once! This has been written exclusively for Experts and is way above your head. The world dominance of the Expert first featured in book form 50 years ago, in this revolutionary work of instruction with an obscure polysyllabic title designed to match its subject, and was so seminal that it has earned a second coming. 'What?' I hear you ejaculate, 'Didn't we have enough last time?' Perhaps so. But that book launched Nonscience—and what it predicted is all coming true, so a new generation of superior beings needs to have access to its powerful revelations. There are some newly written sections (like this) to update the chapters, while the original text now includes present-day values, to show how much money has lost value over the years, and we've made sure that we have gender-neutral pronouns everywhere. Otherwise the original book is republished more or less as it was originally written, half a century ago. How time flies.

The book came out in translation, it continued to crop up in the press, and later became a collector's item. Copies have been selling for hundreds of dollars and the highest price charged for a good copy was $1,500. I'd keep your money, if I were you, and read this reissued version instead.

The original edition of *Nonscience* was translated in a series that included books by Dr. Benjamin Spock, Alex Comfort, Bertrand Russell and Bob Dylan. The book became a collector's item, the most expensive costing $1,500 at Glass Frog Books of Hawthorne, California. I can think of far better things to do with the cash.

Within the top-secret pages of the book you will discover how obscure language and self-referential argument can keep the truth of any subject hidden from everyday folk. You will find out how to patronise the public, how to talk down to them effortlessly and with authority, and how to connive at being given vast and costly apparatus that you do not know how to use, and which performs no useful function, but which (and this is all that matters) your rivals don't yet have. Television commentators and newspaper reporters will fiercely challenge business pundits and regularly humiliate members of parliament by interrupting them angrily in a way seen nowhere else in society, but they never do that if you're an Expert. Then, questions will always be polite, and your answers won't ever be challenged—because nobody understands what you're telling them anyway.

It was science that governed our understanding for centuries. It derived from the Latin *scientia*, knowledge; and by the late 1300s it was being used in English in much the same way as we use it now. But over the last 50 years that has given birth to something far more modern and trendy—Nonscience. Science is based on discovering truth and presenting it to the public. Nonscience is the opposite, for it is founded on falsehood and exaggeration which is kept hidden from the public by incomprehensible constructs and impenetrable language.

BUZZ ABOUT WORDS

Experts always use buzzwords to seem current. It applies now, even more than it did when the original book first appeared. Lasers were becoming popular when *Nonscience* was written, and ever since the Revitalift company renamed their skin cream Laser Renew, sales have gone through the roof (there's nothing to do with lasers in the product). There is an embroidery company called Laser Apparel, and Laser Brand tools … it goes on forever. No product needs any actual connection with lasers, just have the word in the name.

The most commonplace of concepts can be dignified by the use of the right buzzword. Knock down a few trees or plough up some scrub and it interests few; make it a 'loss of biodiversity' and you're straight on television. If you automate some everyday process by computer it's neither here nor there; but use the right software and it transmutes into Artificial Intelligence, guaranteeing you instant headlines. Lay down a road with some traffic lights, and it won't interest anybody. But call it a 'smart highway' and it's straight in the news*—indeed, you can now purchase 'smart tape', and even drink 'smart water'. Add the word 'technology' to

---

* Best not to call it a 'smart motorway' because these have killed dozens of innocent motorists. Were you a mass murderer you'd be in the papers, and in gaol for life; but if you have designed a highway that kills motorists you would be in the clear (especially if you emphasise that 'lessons have been learned'). Whenever people persist, say 'draw a line under it and move on.' That never fails.

something ordinary and it becomes extraordinary: Don't promote, say, a garden brush; make sure it is 'garden brush technology'. If you drop in the occasional 'smart' and 'technology' you can transform anything into a topic commanding awe and respect. Anything: lavatory paper can become 'smart anal hygiene technology'.

Using a buzzword can be powerful, though it can be equally useful to omit a keyword. Many experiments have been carried out on laboratory animals and the results would never apply to humans. So, if you find some extraordinary finding, just omit that it has been found only 'in mice'. This guarantees you headlines every time.

HALF-FULL OR HALF-EMPTY?

You know the old story about the glass that's half-full (if you're an optimist) or half-empty (if you are a pessimist). There are other interpretations. Should you be an engineer, then the glass was the wrong size, while if you're a traditional scientist the glass is completely full anyway (the lower portion is filled with liquid, the upper part full of air). But—if you are an Expert—then this could mark the launch of a new discipline called Containeristic Fluidistical Volumetricationalised Disproportionality with funding running well into six figures and lasting for years. One aspect of Nonscience is taking an idea (something simple—the kind you'd sketch on the back of an envelope in three or four minutes) and expanding it into a decade of research which brings in massive funding. Remember that.

Almost every senior Expert in the world is a man. You'd expect that. Men are more often driven by rivalry and aggression, they take status seriously, and they love to confuse and patronise everybody else in the quest to remain aloof (these days we call it 'mansplaining'). Experts still need everybody else to be subservient to them and their pronouncements; keeping women in their place meant that, without effort, half the population was already under the thumb. It wasn't only men who believed this—women did too. A survey showed that, if you sent out a paper to women referees, sometimes with the author identified as 'John' and at other times as 'Joan', the women referees invariably came back with a more favourable response if they had been told the author was a man. That survey was carried out by Philip Goldberg at Connecticut College when the original *Nonscience* was being written, and when the 1960s era of liberation had flowered to the full. Ah yes, you're thinking, it has all changed since.

Did anything change? Two more American academics repeated the Goldberg study some two decades later and concluded: 'An article written by a male was evaluated more favourably than if the author was not male. Subjects' bias against women was stronger when they believed the author with the initialized

January 2020 kicked off with young readers enjoying this strip cartoon from the *Lego* magazine. These young girls realise that, if they join a group of laboratory researchers, they'll immediately look out of place. They decide to dress like cooks and bring in trays of muffins—such a forward-looking impression for today's young readers.

name was female.' Okay, that was back in the 1980s. Surely it's different now? Don't you believe it: an article published in August 2019, while this update was being written, was entitled 'Committees with implicit biases promote fewer women when they do not believe gender bias exists.' You see? Even when they expressly state that they aren't biased against women, they still are. You cannot believe people aren't misogynistic just because they think they aren't … who controls the TV remote? Who runs the family finances? Who controls the heating? That's the giveaway.

Another notable contribution in 2019 was written by Charles Fox and Timothy Paine, who looked at how the gender of an author influenced the reception of a published paper. They tabulated all their results, and again demonstrated the greater perceived authority of men.

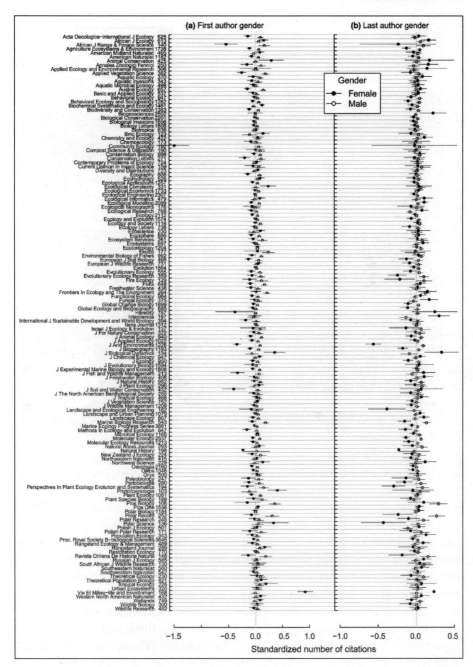

In March 2019, a paper entitled 'Gender differences in peer review outcomes and manuscript impact at six journals of ecology and evolution' included a detailed table showing how papers by men always score higher than those by women. Previous investigations have sometimes relied on small samples, so this time they included more than 23,000 submissions over a six-year period plus 12,000 publications and they list some 50 publications further to prove the point.

Infection with COVID-19 reached Britain in January 2020. The government refused to take it seriously and told people to carry on as usual. Instant action was the only answer, but not until late March did the government instruct public places (bars, restaurants, etc) and schools to close. During those lost weeks tens of thousands of cases were spread across the country. Think of the research revenue this generated, Experts!

We were horrified when the COVID-19 pandemic began. It was nothing of the sort: this was a made-up name. We urgently need a system to name viruses; currently it's free-for-all. Nobody admitted it, but this was actually SARS all over again (the new strain being twice as infectious, though ten times less likely to kill you). For weeks the public were told to carry on as normal, in order to spread out the rate of infection through 'herd immunity', and the Prime Minister appeared on television, cheerfully shaking everyone's hand (and insisting he wasn't going to stop) while Experts were wheeled out to confirm that he was doing the right thing. People died as a consequence, though I didn't notice anyone apologising. Within a few weeks, Boris Johnson was in hospital, fighting for his life. Scientists trudging along silently in the background knew the only course of action was to test contacts and isolate them*, but Experts everywhere said different things, so the government could choose whatever they liked and still say: 'We are follow- ing Expert advice'. When the disease hit Britain, the government traced contacts and had them go into quarantine. This is the only way to stop the spread of a new epidemic. On 11 March 2020 WHO declared it a pandemic, and the next day the tracing stopped and infected people flooded into the country by air and

---

* This was proved to be correct by the experience in Germany. They don't have Experts the way we do, mainly because the German language is already bursting full of long words. Did you know that the German for 'tramstop' is Straßenbahnhaltestelle, or that the German for regulating the labelling of beef is Rindfleischetikettierungsüberwachungsaufgabenübertragungsgesetz? Of course you didn't, but it proves my point. Experts don't stand a chance in Germany, so their government did what I recommended—locating every victim then isolating, diagnosing and treating them promptly. The Germans, as a result, had by far the lowest mortality rates for the new disease (at least five times lower than in the UK).

sea without any controls. The Chinese knew that tracing contacts and isolating them were crucial from the start so they used mobile phone apps which sent automated messages to every contact and, if people didn't stay home, their phone would send a warning. They developed a similar app in South Korea, and used it to control their outbreak. Not until April did we in Britain announce we might create an app to do that, by which time we were all locked up at home while the Chinese cities were once again opening for business. Making everybody stay home (instead of tracking and tracing the infection) was like keeping in the whole school because a couple of miscreants had misbehaved. The virus needed stopping, not the population.

Experts told the people to wash their hands and wear fabric face-masks. They pointed out that people in South Korea wore masks, and had better results at controlling the virus. What they didn't say is that the app was crucial. Since the 1960s people in China had worn face-masks when going out, but the masks didn't stop the pandemic. Virus particles are far smaller than the pores in a cloth face-mask, but we shouldn't let the public realise this.

By April 2020 the highest incidence of COVID-19 was in Newport, South Wales, with just 286 cases per 100,000 population, though nobody knew where they were. In Britain (as we go to press) there are about 3,500,000 people who have had the virus, but nobody knows where they are, so there were 64,319,932 people incarcerated, just because the tracing of contacts had stopped. The government were desperately trying to start it up again, while still trying to seem organised. People travelled hundreds of miles to testing stations, only to be sent back home empty-handed. Nightingale hospitals were set up to meet demand, by moving medical equipment and thousands of beds into exhibition centres. The government kept pretending they had 'built' new hospitals yet hardly anybody was moved in to them, scores were turned away, and—instead of isolating COVID-19 victims and giving them specialist care—they were sent to regular hospitals so the virus spread among patients already ill. People needing urgent treatment were kept away by the sense of danger and at least 10,000 of them died. Scores of thousands suffered devastating long-term side-effects, but nobody was counting those.

Meanwhile there was a national shortage of protective gowns for use by hospital staff. This was because Experts had decreed that gowns were dumped and destroyed after one use. Single-use scrubs are a dreadful waste of resources. Coronaviruses are inactivated at 60°C, so gowns could have been hung in a chamber at that temperature and emerged virus-free. Most scrubs are made of polypropylene, which is permanently hydrophobic (water-repellent) yet the regulations say they should only be re-used three times. They can actually be used for years.

Experts ensure the public are kept in the dark about epidemics. The whole BSE scandal was kept secret for years. In 2016 the NHS staged Operation Cygnus, to model how a fatal epidemic would affect us. The government was 'terrified' at the conclusions. It was all kept secret. Last year a National Security Risk Assessment (NSRA) did it again, concluding that 65,000 people could die and the cost might be £2,350,000,000,000 (£2.35 trillion). They recommended we stock up with equipment (we didn't) so this was kept secret too. For COVID-19 a Strategic Advisory Group of Experts (SAGE) advised the government and they wouldn't even reveal its members. There are other committees of which you haven't heard, including SPI-M, which showed how important quarantine would be if ever we faced a pandemic, and SPI-B, to do with public disorder. These were all kept secret, of course. Now a Joint Biosecurity Centre has taken over from SAGE.

The excitement comes from all the unimaginable inconsistencies we can create. For instance, if you were travelling by ship where there had been COVID-19, you couldn't disembark in case you spread the illness, so you were stuck on board. Arriving by plane, none of this mattered. In the three months before lockdown, 18,120,000 passengers arrived in the UK by air from nations all over the world (including from Wuhan, China). Just 273 were asked to quarantine. It wasn't until 6 June that there were any controls—and then, just for one month, arriving passengers were asked to fill in a form that didn't oblige them to quarantine themselves, after which they plunged·into public transport where they breathed over everyone else. The mayor of London, Sadiq Khan, had restricted the number of underground trains, forcing everybody to pack even tighter than normal, while being seen driven around London in his £100,000 Range Rover. On 3 March on TV he had stated: 'There is no risk in using public transport.' At least 30 bus drivers have since died.

Now you see why Dominic Cummings, who guides our Prime Minister, went on a secret 550-mile jaunt to favourite spots for his wife's birthday, why cabinet minister Robert Jenrick drove 300 miles to see his parents, and their nutty professor Neil Ferguson had a liaison with his married mistress; while New Zealand health minister David Clark popped to the beach, and Canadian Prime Minister Justin Trudeau went to his summer residence for a break. People don't understand— those in authority are appointed to make rules, but not necessarily to observe them. It's not what you do that counts, it's what you tell the public.

You can list topics, mixing myths with reality, and see how easy it is to pick out the spurious subject from sound and solid reality. Consider these: the Amazon as the lungs of the world; the legendary unicorn; single-use plastics poisoning our planet; the Large Hadron Collider opening our eyes to new physics; Crick and Watson discovering DNA; Darwin's unprecedented theory; and mermaids. Can you pick out the ones that have their roots in reality, and list those that don't exist? Of course—the odd ones out are unicorns and mermaids. They have their origins in the real world, whilst the others are all untrue. Welcome to the world of Nonscience. And now, after this topical update … back to the original book from 1971.

CHAPTER 1

Farewell, Science . . .

In an era of rapid change, of self-perpetuating progress towards an undefined and indefinable goal, it is almost inevitable that at some stage technology may overshadow the person that makes it possible: that accepted codes of technological betterment may replace our notions of humanity, and machines rule the minds of men. It is in this state that, both unwittingly and unwillingly, we are beginning to find ourselves.

A glance backward, to an age only just vanished from view, can usefully remind us of the science that was. Society's structure was stratified in a neat, predictable system of horizontal planes. Either you were in one of the layers of this elitist sandwich, or you weren't; and it was damnably hard to move up from one compartment to the next in line.

'Do not try to overreach yourself' was the maxim: 'you will only get out of your depth.'

At the pinnacle of society were the leaders—the gifted ones. Some were concerned with religious or metaphysical matters, while others were embroiled in such earth-bound subjects as the law. But those who were concerned with the behaviour of tangible things, with the nature of humankind and the universe and why materials behave as they do, were without peer. They were—in the traditional sense—scientists.

Scientists were impressive. They dwelt in lofty apartments or in gracious homes with an aura of heterodox bohemianism about everything. Generally bespectacled and frequently balding, they wore severely cut suits of dark and muted cloth and rather jolly silk ties. They were individuals of guarded ways and few words, who stood out away from (and above) the ordinary members of society.

They were sober men with an occasional tendency to light-hearted lechery and knew about simply *everything*. At least, that was their reputation. One might raise his greying head from the well-worn eyepiece of a microscope to observe, with pitiless scorn, that it was obvious to anyone with even meagre intellectual abilities that his house-keeper's twingeing knees that day were no doubt due to electricity in the upper atmosphere. Or to remark that a coin left on the surface of a frozen lake in winter would slowly sink through the solid ice. Perhaps even that a native head-hunter known personally to a second cousin had once used mouldy melon-rind to cure boils … or some such gem of intellectual wizardry.

It seemed you could trust a scientist. All that was needed was an induction coil, several feet of brown string, candles, sealing wax and a microscope or chemical balance to solve almost any problem put before him. He would communicate to his fellow-scientists through publications bearing such terse yet mellifluous titles as:

OBSERVATIONS CONCERNING THE MICROSCOPIC CONFIGURATION OF SESSILE SHINING GLANDS ON *Agrimona odorata* (Gouan) Mill.

or

OLFACTORY AND TASTE SENSATION IN THE MOUTH REGION OF *Viviparus viviparus.*

And he would communicate with lesser mortals rarely, if ever.

In those dim, far-off days (which almost *are* beyond recall) they discovered a rare and precious metal, more elusive than gold, called aluminium. They mused on the physical basis of life, and conjured delicately with the intricate mathematics of crystal formations and magnificent, minute micro-organisms with shells like spun silk. Scientists were great, yet innocuous people. They were trying to tell you more about the world—and they tempered their arguments with a delightful concern for humanity and the sanctity of existence.

They played chess, the violin, billiards or croquet with a precision born of intellect and objectivity; and they were—if a trifle aloof—thoroughly reliable people of integrity.

Science was a decent enough subject, too. It was concerned with facts rather than pointless opinions; with objective reality rather than aimless subjectivity. Basically it tended to concern itself with finding out the truth of a proposition; the facts of the matter. It was not merely a technique for proving that you were right after all. Above all it was concerned with sensible, pragmatic ends: cuckoos and their nesting habits, bees and their dance, the spectrum of the sun and *could* that strange line be a new element? Science was serious, deep, sensible and generally constructive. It was precise, and had even a kind of logic to it. It seemed bent on improving the lot of our species.

The slowly unwinding decades changed all that. Commercialisation stepped in, vanity asserted itself, prestige rather than integrity came to matter above all—and the whole academic fabric altered. With the materialistic vision came a new selfishness, where personal attainments ousted truth, wisdom, propriety and the rest; where the unobtrusive aspirations of yesteryear were swept over by a wave of hasty, destructive, even malignant new 'developments'. The cybernetic simplicity of the scientific attitude was gone, and in its place stood the abstract, almost hysterical striving for 'progress'.

People turned against it, of course. Many of us came to hate what we saw and if we were involved we rebelled. That poses the most irrational notion imaginable—a scientist against science. It is no better than to riot for peace, murder for love or rape for virginity.

So now where are we? Do we really want to abolish technology? Of course not—it has given us more practical benefits than any other of humankind's achievements. Should we simply abandon science? That would be scarcely rational when its fruits have saved so many lives, taught so many hungry minds, fed so many starving people, prevented so very much more suffering than its worst perversions could ever cause. We may choose to quit our homes to live in the fields, to reject artificial methods of agriculture, to destroy our mass media; but we would—with dramatic suddenness—succumb to the effects. That's no answer: a person does not stop eating food to avoid toothache.

In truth it is not science with which we are contending anymore. There is no virtue in being 'anti-science', since the old subject is—simply—not the enemy. There is nothing in the least scientific about this brash new discipline, concerned with topicality, financial viability, sensationalism and wealth. Neither is it a new form of science—indeed, in most respects it is the very antithesis of that traditional profession.

On the contrary, it tends to be typically obscure, unrealistic, pointless, harmful, incomprehensible, thoughtless, short-sighted and unrelentingly inhumane. Science was never like that. No, it is something else, something new, something quite, quite different. We have witnessed the birth—of *Nonscience*.

It is Nonscience which, alone amongst the many factors that influence the inhabitants of Britain, was able to authorise a simple change of time standard that drove the entire population magically from their beds in the pitch blackness of the depths of winter. Not even a concerted raid by the Luftwaffe in the Second World War could be relied upon to do that. Only Nonscience could subject millions to the regular, unannounced sonic explosions of the supersonic transport aircraft era and hope to get away with it. Nothing else in the world could have made gasmen do away with the therm and adopt in its place the gigajoule, or convince traffic authorities that we ought to abandon miles an hour in favour of metres per second (which makes an approximate calculation of the time taken to drive from A to B into a major exercise of higher mathematical expertise). What else but Nonscience could have conspired to make microscopists throughout the globe reject the time-honoured unit of measurement *mu* (written µ, as I am sure you knew all the time) in favour of its new name, the micrometre (not to be confused with the micrometer, of course, which is something entirely different)?

What but Nonscience could possibly allow the entire economic future of the United States to be based on a multivariable model for gross national product, a model from which any statistician would instinctively recoil with emotions not unlike panic and horror? And what else could ever have reiterated, with banal monotony, that each moon probe, each Apollo shot, each new experiment, was going to 'unravel the mystery of the origins of the solar system' when in fact we *now* know no more about the lunar craters than we did over 300 years ago? Only Nonscience, pragmatic, self-fulfilling narcissistic Nonscience, could so blatantly alter our environment, presume on our tolerance, decimate our lives and upset our personal well-being in so many ways simultaneously—let no-one confuse that with science!

Why Nonscience? I coined the term for basic etymological reasons since— like all the best words (but quite unlike most of those in Nonscience itself)—it is virtually self-defining. *Non* is a prefix used to denote *not*, and *science* is a word meaning, well, *science*; and no-one has come forth with a reasonable definition of that for years.

Next, how do you pronounce it?

Like all the best terms of Nonscience itself, there are several choices. The basic pronunciation is as it stands: *non science*. In this form (which is how Americans prefer to hear it) it means exactly what it says; the implications of it are clear, yet can be contrasted clearly and unequivocally with 'science', which is just as well.

A refinement is to adulterate the *science* until it sounds like *see-ence*. The term said thus acquires mystical, almost oriental overtones. It has a slightly European flavour to it, and several Americans I know think it sounds cultured that way. The meaning is a little less clear when the term is pronounced in this manner, and that can be a pronounced advantage.

And the third derivative is to rhyme it with *conscience*. There is an implied contrast between this word and Nonscience, which this form of pronunciation subtly reveals; and at the same time the term sounds a little like *nonsense* which allows further subliminal implications to become manifest. It sounds, in this way, like *non-shunce*.

Just how you say it depends on the company you choose. The exact form depends on whether you wish to be understood (most practitioners of the subject do not, as a rule) and just whose side you are on.

I always pronounce it 'Nonscience', actually.

What of the perpetrators of this new subject? Clearly, since their aspirations are so very far from those of the earlier men of science to whom we have already referred, they can hardly be called scientists. Nor, for that matter, are they non-scientists either, for they embody many of the superficial traits peculiar to scientists in the traditional sense. They have the same authority, the same status, the same bearing; and they have acquired the same responsibilities. But they gain these attributes by entirely different means.

They utilise the mass media, for one thing; by posturing in the public view, on television and in the newspapers and magazines, they cease to be laboratory-bound specialists, and become household pundits like the bank manager in your closet—always available, always familiar, always ostensibly there to help; yet totalitarian in their power and able to make pronouncements against which there is no appeal.

And underlying the whole trend is a dichotomous tendency to specialisation, so that vertical fissures appear in society to replace the horizontal strata that went before. Specialists in one subject are competent only in that. They will discuss nothing else, giving an opinion on naught but their own specialised field—and yet they will, by their actions, presume to interfere with all of us if they deem it necessary. If their own research, their particular concept of 'progress' is to press ahead then (with no social responsibility implied or requested, even) the insularity disappears and wholesale alterations of our whole tenor of life can well ensue.

This is a new breed of individual. Unquestioned, infallible, omniscient; trusted above all to guide lesser mortals along life's precarious path.

They are—the Experts.

Think of the aura attached to this expression. The expert witness who alone amongst people can give an opinion to a court of law and find it valued as highly as the currency of truth, the expert opinion that gives a hard-and-fast yes-or-no decision on matters of importance in everyday life ... already we have begun to accept the existence of this breed of super-person. Now at last we can grant them the status of an officially recognised proper noun, and we may elevate that initial 'e' into a capital letter, as humankind has done for their peers, masters and lords for centuries past.

Experts are no ordinary mortals. They are dedicated to data. Facts and figures can be made, by the Expert, to substantiate any claim. Indeed a truly devout Expert can use identical data, on different occasions, to substantiate entirely opposite points of view. Experts are opinionated, self-centred and frequently irrational. They are trained to obey the dictates of the establishment without question, and to ignore such considerations as would put their discipline into disfavour. They are selected by a process that ensures conformity and eliminates with ruthless efficiency any real talent for originality.

Scientists of old would work on subjects of common interest, or which were likely to throw up discoveries for the benefit of humankind; the Expert of today is concerned primarily with what is newsworthy, and what is in the public eye—what, to coin a phrase, is *Fashionistic*.

This is an important new concept for us to grasp if we are to understand the

> Voiceprints represent the human voice pictorially and are said to be as unique to each individual as fingerprints.

> sound spectrograms, unlike fingerprints, are subject to considerable variation; a person's voice changes with age and circumstance and can be disguised or made to resemble that of another person.

> There is no acceptable evidence to support manufacturers' claims that the voiceprints recorded by their equipment identify an individual almost as uniquely as do fingerprints.

What Fashionism failure did to the voiceprint.

way that Experts function. In 1971 the most recent chart-toppers in the world of fashionism have been the pill, women's lib, marijuana; some years back it was voiceprints, fallout and food additives. Then it was automation, currently it's the environment; in the 1960s it was cholesterol, in 1970 by was DDT. For the Expert it is a race against topicality.

And behind it all is the use of particular terms as a language of Nonscience, a language that only Experts can fully understand. Phrases such as 'design tolerance', 'mean survival rate' and 'experimental error' crop up with predictable words including 'parameter', 'programmed', 'notional' and the like. Not only that, but Experts have a positive penchant for the coining of new terms that mean little enough to them, and nothing whatever to anybody else. Possible examples might be 'psychogenetimorphic' or 'micro-organismically reorientational', and the title of this book is no more than an example of what we may soon be in for. Indeed no self-respecting Experts would be content with a term like Fashionism; to them it would have to be something far more intricate and Nonscience-orientated—*quasi-notional fashionistic normativity* is more like it …

Textbooks for Nonscience are, until now, lacking. And this brings us to the crux of the argument: how does one become an Expert? If Nonscience is to continue to proliferate and expand in the future (as it undeniably will) then how do you get onto the bandwagon before it is too late? That is what this book is all about.

It is a study of the working of Nonscience and the transformations (pseudo-transmogrificationalific egocentrified reorientational proclivities, to be precise) that are possible in adjusting one's brain to it—a process that an Expert would prefer to call expertistical cerebrointellectualised redeploymentation. We consider the training aspects, the financial question of a career as an Expert and look also at the professional prospects that result. Naturally we touch on the practical implications for all humankind.

Some people (figures currently available suggest that it is a very small minority) prefer to remember science as it was, and claim to recognise hazards—even disaster—in the present state of affairs. It is not for me to side with a minority view, now is it? And so, in the best traditions of social democracy, this book is forward-looking, thrustful, progressive; the contents are a handbook for aspiring Experts—and, mark my words, if they follow the advice herein they cannot but succeed!

The faint-hearted had best read no further. For these men and women a life of seclusion is the only hope. But for the student of Nonscience—as the following chapters show—cultivation of the correct attitudes, adoption of the right opinions and strict adherence to the professional code can, between them, assure anyone of a rosy and profitable future.

Almost anyone, at any rate …

———————— —m— ————————

EXPERTS RULE

Fifty years have passed since those words were written and, in the 21st century, Experts really are ruling the world. The public don't get a look in. The media are baffled, governments hopelessly lost, grant-giving organisations gullible, and beneath it all lies a whole raft of predatory publications and conferences with spurious titles that cream off vast profits for doing very little. As an Expert, you are unassailable, because nobody else can understand anything you say. You can travel on expenses to spend a week in a luxurious spa in Hawaii or go on a luxury cruise around the South Pacific where you will learn little and impart less; but where the nightlife is sensational and drinks are cheap. Everybody knows what is going to be said at a conference, and anybody with a commercial aim or a private project for which they have funding isn't going to reveal anything new. Their employer won't let them. Everybody knows that. Experts with a funded project have ultimate power, because they will be given vast sums of money to indulge their whims—but only so long as nobody else understands what those whims might be.

Continuing Medical Education (CME) in French Polynesia ───────

1 ───────────

SOUTH PACIFIC CRUISE, 2021 — MEDICAL CBT FOR ANXIETY: TEN-MINUTE COGNITIVE BEHAVIOR THERAPY TECHNIQUES FOR REAL DOCTORS
27 Mar 2021 - 07 Apr 2021 • Papeete, French Polynesia

Organizer: CBT Canada

Abstract: CBT Canada is pleased to offer a CME cruise from March 27 to April 7, 2021 (11 nights) aboard the spectacular m/s Paul Gauguin, the #1 Mid-Sized Cruise Ship in the World for 2019 (Travel+Leisure). The CME is the acclaimed "triple strength" CBT for Anxiety module, accredited by the College of Family Physicians of Canada (you'll earn 36.0 Mainpro+ credits in 12.0 workshop hours'). Join us exploring the South Pacific and you'll gain a marvellous set of ultra-brief techniques to help patients overcome shyness, worry, panic & other common anxiety disorders. Head instructor Greg Dubord, MD is nice & limber from giving over 500 CBT workshops, and a University of Toronto CME Teacher of the Year. Assistant faculty includes Christine Kennedy, MD, FCFP, FRCPC, who will be reviewing recent studies on the objective diagnosis of psychiatric disorders. 'American Academy of Family Physicians (AAFP) members are eligible to receive up to 36.0 Prescribed credit hours for attendance at CBT Canada's 12.0 hour (/36.0 Mainpro+ credit) workshops due to a reciprocal agreement with the College of Family Physicians (AAFP, 2016).

Contact: CBT Canada;   Phone: [8774668228];   Email: info@cbt.ca

Topics: CBT, cognitive behavior therapy, CBT Canada, continuing medical education, anxiety, fear, phobias, shyness, worry, panic, tools

Event listing ID: 1290077

Related subject(s): Psychology

Event website: http://cbt.ca/locations/cbt-south-pacific/

Experts can travel almost anywhere they like, so long as they are attending a conference. Many former travel agencies now describe themselves as conference organisers, and make greatly increased profits. Here is a luxury cruise that has been made to look like an academic conference, with points to be earned. The difference between this and a holiday is that your Department pays the bill.

The failure of the public to grasp realities is evident around us. People are convinced that DNA (which was actually discovered by the Swiss biochemist Fritz Miescher in 1869, and not by Crick and Watson in 1953) is solving a host of our medical problems. It isn't. The decoding of the human genome has been celebrated as a triumph for the Human Genome Project, costing about $3,000,000,000 (3 billion dollars). They don't like to reveal that the genome was decoded much quicker by an independent investigator Craig Venter at a cost of just $100,000,000 (100 million, one-thirtieth of the cost). Indeed, Venter was racing ahead so fast— and so cheaply—that he caused the Human Genome Project to speed up and finish quicker than it was planning to do. This reminds us that Nonscience must be protected. Speed is never an issue. Costs should be maximised, and never contained. And you must always keep the truth from the public. At the time of the DNA project, the public were bombarded with promises that a new era would emerge where recalcitrant diseases would be banished, and a new era of public health was about to dawn. Little has happened since, of course; but we keep quiet about that, and we never mention the waste of billions of dollars. And the one thing we hate the public to realise is that the geneticists decoded only 2% of the genes—98% of our DNA remained a mystery. Nobody knows what it's there for, and nobody knows why we have it. Not only was the Human Genome Project 30 times more expensive than it need have been, but it didn't reveal what the public were told it revealed, it only listed one-fiftieth of what was there, it took far longer (and cost far more) than was necessary, and it didn't provide the results that were promised. Keep this under your hat.

ASTRONOMY MYTHS

Astronomy is an object lesson. There are huge sums of money to be had, people will believe anything you say, and nothing you do has the slightest effect on how people live their daily lives, so the possibilities are endless. You will have countless pictures which can be exaggerated anyway you want, and tons* of data that can be used to create endless myths. If you increase the hue intensity of digital images you can obtain pretty pictures, which people don't understand but which would make very nice album covers or wallpaper and look beautiful on television. What you mustn't admit is that we don't understand, and cannot even detect, most of the universe. If you work out how much stuff is out there, then you immediately realise that we are aware of only 15% of it. The remaining 85% we call 'dark matter' and nothing more is said. The fact that our basic grasp of space is so feeble is something we try to keep from the public, for it is far more rewarding to keep endlessly repeating what we do know. On the other hand, there are compensations; this is a field in which you can simply invent stuff to suit yourself. You can make up a discovery which actually means nothing, and which exists only in your imagination, for nobody ever questions it. You can claim huge headlines for fiction!

Monica Grady, a professor of Planetary and Space Science, came up with one such idea in February 2020. She said that it is 'almost certain' that there is intelligent life on Jupiter's satellite Europa. There would be, said reports, octopus-like creatures. Between you and me, there isn't a scintilla of evidence for this absurd claim. The coldest parts of Europa never go above –220 degrees Celsius; even on the equator the highest temperature never exceeds –160 degrees Celsius. Of course it's crazy – but it came from an Expert, and made plenty of headlines. Grady is a professor at Liverpool Hope University, which you probably have never heard of. It scores very low on the Complete University Guide web site, with only one star (out of five) for research, for academic research spend and also for good honours degrees. However it scores five out of five for student satisfaction, just as you'd expect.

For example, in recent decades people have noticed that the light intensity from distant stars fluctuates slightly. You'd be right to show surprise that this hadn't been done before, since astronomers have been minutely observing and recording these stars for centuries and have been able to monitor brightness for decades. In any event, you could account for fluctuations in brightness if you postulate that a planet is orbiting the star, just as we orbit our Sun, and so the idea of an 'exoplanet' was born—a planet orbiting a star millions of light-years away, and therefore unimaginably remote. That's how a traditional scientist would look at it. But in the era of Nonscience that changes dramatically. Instead of

---

* Oops. *Tonnes*, sorry.

Monday, Feb 10th 2020 12PM 7°C 3PM 8°C 5-Day Forecast

Mail Online

Science & Tech

Home | News | U.S. | Sport | TV&Showbiz | Australia | Femail | Health | Science | Money | Video | Travel | DailyMailTV | Discounts

Latest Headlines   Login

British scientist says it is 'almost a racing certain' that the icy seas on Jupiter's moon Europa are home to alien life that are 'octopus-like creatures'

- Monica Grady believes it is almost certain that Europa is harboring life
- However, she believes they are octopus-like creatures living in the icy seas
- Grady thinks the caverns and caves on Mars may also be housing life-forms

By STACY LIBERATORE FOR DAILYMAIL.COM
PUBLISHED: 18:18, 7 February 2020 | UPDATED: 19:41, 7 February 2020

Share   6k shares   1.3k View comments

Monica Grady (pictured) is a professor of Planetary and Space Science at Liverpool Hope and says that by looking at the bigger, inter-planetary picture, Earth's own ecological situation is brought into sharp focus

A British space scientist says it is 'almost a racing certain' that Jupiter's moon Europa is home to alien life, but believes they are 'octopus' like creatures.

Monica Grady, who is a professor of Planetary and Space Science at Liverpool Hope University, suggests the icy seas beneath Euorpa's surface is a prime location to find beings with similar intelligence to the marine animal.

Brady also thinks that the deep caverns and caves on Mars may also be harboring life-forms, as these areas provide relief from the intense solar radiation.

pants

TRY TODAY

In February 2020 the ridiculous idea of octopus-like creatures living on Jupiter's moon Europa was promoted by a professor at a minor university in Liverpool. There isn't the slightest suggestion this could possibly be true and no evidence for it (the highest temperature ever recorded on that satellite is –160°C) but look at the headlines! This is the way to go.

postulating an orbiting body that may (or may not) exist, you invent one of your own. Just like the tellers of fairy tales, you create an imaginary world. You give it the hint of clouds, an alluring misty atmosphere, perhaps the outline of continents … just invent something out of your head. This is like drawing Snow White's forest or Never-Never Land. At this stage you will probably need a graphic artist to immortalise your dreamy speculation in a format that's suitable for publication. Once you have your image, release it to the press and they will all devour it with eagerness.

It will feature in magazines, in newspapers, and all over the Internet. We both know that it's pure invention, made up like a fairy tale, but the media won't question its authenticity. Many of the pictures are released in the United States by the National Aeronautics and Space Administration (NASA) and people trust them, because NASA are well known as the world leader in space.*

Once in a while, somebody will raise a query about the wisdom of all this lavish expenditure. In 2019 the New Zealand government agreed to donate NZ$25,000,000,000 (about 17 billion American dollars) towards a new telescope. Those in the field rubbed their hands with glee. But Professor Richard Easther, the most senior astrophysicist in Auckland University, could see it was a waste of money. He concluded that it was ridiculously expensive and had been massively over-hyped. Now, truth to tell, all these schemes are exactly that. The difference here is that Dr Easther said so! Said he: 'My own belief is that academics have an obligation to … speak out about matters of real public importance.' That's unforgiveable. Worse, 11 other leading New Zealand astronomers signed a letter saying he was right.

The establishment turned on him with fury. In any other field they'd be explaining how they felt the costs were worthwhile, they would list the benefits, and provide some cost–benefit analysis to help change his mind. That's how it works in the real world, but not in high-budget astrophysics. Easther was berated, personal attacks showered down on him, insulting messages were circulated, and one journalist (who wanted to investigate the controversy) was warned: 'Be prepared to have the finger pointed at you if and when mental health issues get dragged out in the media.' There is a lesson here for everyone: query the new television that your spouse wants in the living room any time you want. But: question the wisdom of a government planning to sink billions of dollars in some dubious project that benefits no-one? I wouldn't, if I were you.

Experts exaggerate everything to do with the skies. They will predict huge meteor showers, knowing few people will go to watch (just as well, since the most they are likely to see is a few quick, tiny sparks every hour). Images of meteor showers always show the sky covered with shooting stars; this is fake news.

---

* Strictly speaking, America lost the space race from the start. It was Russia who launched the first satellite, sent the first signals back from space, launched the first animal into orbit and the first lunar spacecraft, sent the first images from the far side of the moon, returned the first animals (and plants) from space, put the first astronaut in orbit, sent the first spacecraft to Venus, launched the first dual-crewed spacecraft, put the first woman (a civilian) in space, made the first spacewalk and the first soft landing on the moon, and was the first to send back images from the surface; first to send a probe to Venus, first to perform spacecraft docking, and first to fly astronauts to the moon, orbit, and return; first to return to Earth a sample of moon rock, land the first lunar rover, achieve the first soft-landing on Venus, and send back the first signals from another planet; to launch the first orbital space station, first to land a spacecraft on Mars to send back data (50 years before the 2020 missions) and the first to launch a multinational space station crew. Then, ever since NASA's space shuttle failed its Health and Safety certificate, the United States has relied on Russian expertise to launch their American astronauts. America was rightly applauded for launching astronauts into space, though less is said about the Redstone rocket; it was not an American rocket at all, but a souped-up Nazi V-2 designed and constructed by the enemy rocketeer, Wernher von Braun. He was supposed to be tried as a war criminal, but had his military history wiped clean when he volunteered to work for the American enemy instead. Designing the Apollo rocket that put Americans on the moon was his greatest triumph. You see? With keen determination, unwavering secrecy, and plenty of professional PR, you can claim anything.

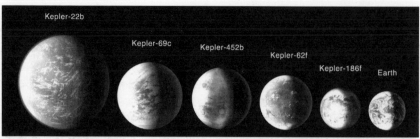

← The chance to publish pure invention awaits you in the field of astronomy. Like UFOs and Science Fiction, it is all invention, and nothing of what you say has any relevance to the people who hear your words, or to the problems our planet faces. None of it can be proved true or false so you are able to say virtually anything you like. You can even invent non-existent worlds. Vivid guesswork gives us concoctions like the Alpha Centauri system dreamt up by R D Nickel (*bottom*), and the two invented imaginings by NASA (*above*). All these images are pure guesswork, a kind of fairy tale for the gullible masses, and there can never be any comeback because we will never be able to know the truth. Just one of those planets is real (can you guess which one?).

Typical portrayals of meteor showers show massive fireballs descending in torrents. The above example (*top*) is modest in comparison with some I've seen. Since the public have never seen anything remotely like this, they naturally consider the Experts to be far better at observing. A real shooting star is more like the other example (*bottom*). This one is much brighter than average, but—if you wait five minutes or more—there might be another one. If you're lucky.

The aurora is another case in point. Every time the public see a movie showing the aurora borealis it is bright and brilliant, flickering and pulsating, like luminous ferrets fighting in a pair of tights. The real aurora is nothing like that. It hangs pale, faint and listless, high in the heavens, slowly changing and sometimes drifting like a diaphanous curtain in the slightest breeze. Your eyes need to become adjusted to the dark to see it properly. But not when it's on television! By then it has been speeded up to frenetic activity in time-lapse, and image intensification has increased the soft subtlety to dazzling, vibrant colours. And this reminds us of a crucial truth: it is the public perception that matters, not the reality.

WORLD'S LARGEST

If you want the ultimate example of the power of PR, then it's the Large Hadron Collider you need. This is the world's biggest machine, the most costly single experiment in the world, one of the largest employers, and—to top it off—it has given us no benefits at all. Powerful colliders have been built for decades in the search for fusion energy (like the hydrogen bomb). The first accelerator was built in 1928 in Germany, and others followed in the 1930s. By 1951 a thermonuclear reactor called Huemul had been built in Argentina and they soon announced that they had achieved fusion. They hadn't, of course. In the 1950s at Harwell there was Zeta, standing for Zero Energy Thermonuclear Assembly, which promised unlimited power. It wasn't zero energy, and it didn't provide thermonuclear power, but it still made headlines for years and in 1958 they announced they'd achieved fusion. They hadn't. Meanwhile, a reactor in America in 1957 almost gave the game away; it was called the Perhapsatron Device. That's much too honest. Fusion power certainly existed; it gave rise to the world's most powerful explosion ever, when the Russians detonated their 50-megaton Tsar Bomba in 1961. But it could not be harnessed to make electricity.

That's how matters stood when the original *Nonscience* was published. But I predicted that there would be 'opportunities for those who wished to work with subatomic particles' and that came true with a vengeance, when plans for the Large Hadron Collider (LHC) were announced. They said it would provide answers beyond the Standard Model of physics. And has it? No. Are we nearer a new understanding of physics? Of course not. Theoretical physicists had listed new subatomic particles that they knew the LHC would reveal—and has it? Certainly not. Not one of them. Do we know more about anything that we didn't know before? Yes! We know how to lift billions of dollars from the public purse and spend it without anybody being any the wiser—which is the ultimate aim of Nonscience. The LHC cost almost $5,000,000,000 (five billion dollars) to build and was launched in a blaze of international publicity. They nicknamed it the 'Big Bang Machine' and it was on radio and TV, in every newspaper and magazine, indeed the BBC even had a play written about it. It was a public relations sensation. Although it had not been soldered together properly and stopped working when it broke down the following week, the world knew about its launch. It still costs $1,000,000,000 (a billion dollars) a year to operate and to pay the 2,500 permanent staff. They host about 18,000 visiting physicists each year and there are 12,500 scientists of 110 nationalities from 74 countries 'analysing the data'. They claimed to find the Higgs Boson,* which everybody knows about, though hardly anybody understands, but they didn't really. All they did was detect some activity that was compatible with the mysterious Boson, and the physicists rushed to publish hundreds of research papers, but it was just a weird effect and nothing to do with their strange particle. Instead, they named it 'The God Particle', yet another absurd exaggeration everybody accepted without demur. How do people get away with it?

Meanwhile, although a whole range of new particles had been predicted, not one of them was found. Christopher White, a theoretical physicist at Queen Mary University of London, reminded people that: 'Searches for extra dimensions and black holes at the LHC have been going on since the 2000s.' That is true. But not one has ever been detected. Yes, the LHC has been a colossal consumer of money. People have to understand: if it's paying Experts to indulge themselves while their schoolmates fritter away their lives in everyday careers, it can never be a waste. This is what Nonscience is all about. The *Financial Times* quote Lyn Evans, the project leader, saying: 'The only way we justify our existence is to build more and more powerful machines,' so they've just announced a new one that's 100 km in diameter costing £19,000,000,000 (£19 billion). It might find more Higgs Bosons, and fleece the public still more. And the quest for fusion power? The French, with backers including the Chinese and Russia, are building the International Thermonuclear Experimental Reactor (ITER). It is just 98 metres in diameter and they're confident it will do the job.

If you ran a business like the LHC that failed so conspicuously to do what you had stated when you were claiming funds, you would be fired, gaoled, or end up in court. This reminds us of a cardinal principle: as an Expert, you are above the law. There is never any comeback if things go wrong. When disasters happen, all you have to say is 'lessons will be learnt' and you're completely in the clear. People used to make sure things were safe for the public; not anymore. All you have to do nowadays is to make sure that the cost of compensating the public who suffer is less than the cost of making things safe. That keeps your employer happy. If you're an Expert, even in the worst-case scenario, you may have to move to

---

* The Higgs Boson is named for Professor Peter Higgs who wrote about it in 1964, based on ideas by Japanese-born theorist Yoichiro Nambu at the University of Chicago. The particle must exist—if you work out the mass of all the particles we know, then you find that matter actually weighs far more than it should. So there must surely be a particle that confers the mass. It's obvious, but don't tell people you know.

another post: Every time there is a catastrophe, even if people die, all you have to say is that 'lessons have been learnt' and nobody can touch you. In India in December 1984, half a million people were poisoned and 20,000 died when a Union Carbide Corporation plant in Bhopal released methyl isocyanate into the air. This was the greatest mass poisoning in history. Dozens of legal actions on behalf of those who had suffered were launched in the United States, but nobody understood anything so (of course) the actions were all dismissed or redirected to India. Not until 2010 was anybody brought to justice, when several Indian employees were sentenced to two years gaol and a small fine. Don't worry; they were all free after a few weeks.

You may recall the thalidomide episode back in the sixties. Thalidomide was a drug developed by former Nazi experimenters which caused deformities in 20,000 infants around the world. Those responsible were never punished, indeed (though you aren't supposed to know this) the drug is still prescribed in Latin America, with deformities often being the result. Then there was the poisoning of pregnant women in Corby, who were contaminated by toxic dust from an old steelworks, and whose babies were born with deformed hands. The council responsible had to pay out almost £1,000,000 to each family, but still nobody was ever censured for what they had done. It really is worth being an Expert; you can have all the fun you want and literally get away with murder.

A current example is that of perfluorooctanoic acid (usually known as C8) produced by the DuPont company. You will have this poisonous chemical inside you as you read these words. C8 first came to light when cows near the Dupont factory started dropping dead and it has since been linked to ulcerative colitis, thyroid disease, raised cholesterol, pregnancy-induced hypertension, kidney and testicular cancer, and several other diseases. It is in the soil, in water, and in the air you breathe, and it is highly persistent. C8 may be the most pervasive poisonous chemical in the world. The company was eventually prosecuted—but, Experts, you can breathe a sigh of relief. DuPont were forced to pay $761,000,000 (over three-quarters of a billion dollars) compensation—but they never admitted they were to blame, and not a single person was censured.* Here too, Expert pronouncements prevailed, even though thousands of people had been poisoned and our environment remains permanently polluted.

The fate of the Grenfell Tower residents who burned to death in London was sealed when Expert advice told them to stay in their apartments and not try to escape, while the fire brigade knew they could not rescue them. Fire extinguishers in the building were old and outdated, the emergency lighting was faulty, and the

---

* There has since been a cinema film made about this episode, released in February 2020, entitled Dark Waters. It tells the entire story of the DuPont company and how they managed to escape from censure. Whatever you do, avoid this movie. It is riddled with negative propaganda. If people ask about it, tell them to watch the 2002 film Dark Water instead; that's a horror film and much easier to watch.

cladding on the building was inflammable (and fitted against regulations), and as a result 72 residents died. Kensington and Chelsea Council was later fined £120,000 ... for a breach of data confidentiality. The chief of the brigade, Dany Cotton, paid the price by retiring four months early on a full pension, with music playing and a farewell parade as she left.

A WORLD FREE OF BLAME

Truth was hidden from the public by a barrage of Expert statements when the banks collapsed in 2008. The British government handed £500,000,000,000 (half a trillion pounds) to the banks, who had run into difficulties caused entirely by their own duplicity. The cost to American banks was astronomical—over $2,000,000,000,000 (two trillion dollars). I am proud to report that the banks steadfastly continued in the traditions of Expertism, inventing terms like Quantitative Easing (meaning 'printing more money') and acting as though it wasn't their fault. Thousands died—the crash led to 5,000 more suicides being recorded than normal. So vast was the crisis they caused that the banks knew some of their people must be prosecuted, and this time they were. A grand total of 13 bankers were charged with fraud. The majority were acquitted and most of the others earned a few years in gaol (a minute price to pay for the vast sums they'd made). The banks, meanwhile, carried on as they were, this time selling £53,000,000,000 (53 billion pounds) worth of Payment Protection Insurance, often without telling their customers they had been sold anything at all (the banks just took the money from the customers' private accounts).

A current example is the Wells Fargo Bank. After the financial collapse of 2008, they emerged relatively unscathed and with a sound and solid reputation. They capitalized on this by a massive fraud, setting up non-existent accounts for customers who didn't know anything about these new arrangements. At its height there were 1,500,000 fictitious accounts and over 500,000 fraudulent credit cards set up by the bank. They set each credit card PIN to 0000 so the bank had complete control over the customer's financial affairs, without the hapless client finding out. Well, eventually they did; the names of the fraudulent bankers soon came to light, and the chain of command was revealed. Eventually their CEO, John Stumpf, paid the ultimate price by agreeing to 'step down'. Since then Wells Fargo have been signing unwitting customers up for unnecessary auto insurance policies and have been fined another $1 billion. The Experts behind it all? No penalties for them. It was the same story in February 2020 when three top Barclays Bank executives were prosecuted for fraud after a 7½-year investigation. All were acquitted.

If there is ever a situation where fraud can be demonstrated, the banks—not the executives!—are fined some money, which they reclaim from their charges to customers. It is a brilliant ruse, and the same bank officials are still raking in unimaginable bonuses and vast salaries.*

It was the same with the E-Borders scheme intended to check arriving visitors. It never worked and cost £742,000,000 to write off. The Firecontrol project was going to link up 46 fire stations, but that failed too; it cost £470,000,000. Then there was the National Health Service IT system. Slow, inefficient, riddled with bugs, it was a failure from the start. One contractor, Computer Sciences Corporation (CSC), reportedly never supplied what it promised. Another, Accenture, was said to have found it too hard to handle and just walked away. The total cost to the public was £10,000,000,000 (ten billion pounds).

It was worse when Post Office Ltd introduced the Horizon IT System to take command of the accounting. This time many people lost their livelihood, and some went to prison. By 2013 an internal report confirmed that the system was clearly not fit for purpose. Naturally, this was kept secret. Those enquiring reported that the Post Office refused to hand over files and meanwhile people working for them were losing their homes and their reputations. Jo Hamilton was one manager who was told she had mislaid £36,000 and was eventually threatened with 14 charges of false accounting. Sarah Boyd lost all her life savings. Wendy Buffrey was accused of making off with £26,000 and Balvinder Gill was falsely accused of stealing £108,000. Eventually, Post Office workers shared £58,000,000 in compensation.

Meanwhile, did any of the people who caused the problems suffer any kind of penalty? Don't be silly. This is the age of Nonscience. Experts don't get penalised when things go wrong. They just get paid.

It is ordinary members of the public who are taken to court when they do wrong. A disabled and penniless pensioner was recently taken off to gaol for not buying her television licence. I am just reading about a starving, impecunious woman being prosecuted in the courts for taking an outdated sandwich from a refuse bin at the back of a cafe. She should have read this book. Then she could claim she was conducting research into the 'Psychological Responses to Moisture Retention of Prepackaged Comestibles Retrieved from a Recycling Facility' and gone on a free cruise to Tahiti. Some people never learn.

---

* I had a far better idea. The money handed out to the banks was the equivalent of £850 per head of the British population. I think it should have been released to the people instead. Given a cheque for £850, they would all have put the money—straight in the bank! So the banks would have had the same injection of cash, and meanwhile there would have been an unprecedented boost to the economy with everyone having a large sum of money to spend (or save) just as they wished. Could this have been done? Of course it couldn't—that's not how Expertism works. You need to keep the public at bay and delude them, not do them favours. Once that started, there'd be no end to it.

CHAPTER 2

What is Nonscience?

The recent establishment and expansion of the Nonscience Movement is a matter of great consequence for humankind. For too long we had been bogged down by outdated concepts of thought and behaviour, and the ordinary Lay People of the Western world have clearly been looking for a spiritual leader—a figurehead of responsibility and strength on whom they could lean in times of uncertainty or stress. Indeed in 1970 the mushrooming spread of arthritis, coronary artery disease, cancer, strokes, and ulcers (the 'diseases of civilisation') is mainly due to this feeling of insecurity amongst the general public, simple souls that they are.

The fading of the old religions, the dying principles of family loyalty and tradition, have left a void that simply had to be filled. And now, at long last, it is possible to announce the new movement that is taking over these very roles. It is a movement of strength, modernity and progress; a symbolic fusion of technology, memorisation and the communicative powers of the mass media; a new creed for tomorrow's citizens.

And as any faith must have its disciples, so Nonscience is begetting its own messiahs. They are people of discipline, training and the highest social significance. It is they, alone amongst the world's varied and infinite human population, who pass judgement on matters too weighty for ordinary Lay Folk to understand; they alone who may—without direct attribution—be cited as the originators of schemes designed to alter our entire way of life, our whole environment. Upright, respected, peerless and towering above their neighbours in influence and ability, it is on them that the hopes of the future are rested. They are—the Experts.

No longer need we be confused by the machinations of official departments, traditional scientists in waistcoats and spectacles, rival factions of conflicting interests arguing in public over complex issues that do not concern Ordinary Folk. For in today's world, where the need to get Expert advice is at a premium, Experts hold sway—and in them may be vested the responsibility to make the Official Pronouncement that we must all accept. In them alone lies our hope of peace, security and trust.

AS THE OLD RELIGION DIES,
THE NEW TAKES OVER
… any faith must have its disciples.

The work of Experts is already giving some new, updated meaning to the old-fashioned, traditional edifices of a bygone era. The cathedral of old is reborn through Nonscience—an infusion of modernity and progress that gives worship a new meaning.

A brief look backwards* will demonstrate how badly needed is such a new approach to the subject.

In the past the old-fashioned practical scientists used several quaint, outdated criteria by which to judge their work. Let us extricate some of these obsolete concepts from the obscuring mists of history for a moment, so that the scintillating light of contemporary Nonscience can show them in their true colours—pathetic pastel notions that they were.

---

* Something the Expert does not, in the ordinary way, even contemplate.

INTELLECT

Intellect is the faculty of reasoning, the empty ability to 'think'. Now that may be all very well in certain circumstances, but it's all subjective when you get down to it. What we need for our crisp new technological age—the era of Nonscience—is something altogether more repeatable; something we can quantify a little more. It is important to know that such-and-such a phenomenon occurs whether we are in Mongolia or the plains of New Zealand; in short we need a set of hard-and-fast immutable rules to memorise. That way we know exactly what we are doing.

For this reason we find that the mainstay of modern university training is not intellect anymore, but memory. Aspiring Experts are given a set of pre-digested, packaged facts to assimilate. They are instructed in a predetermined, precise manner and the outcome is predictable instead of haphazard. It is a vast improvement on the hazy, indeterminate aims of yesteryear. Whereas we would once have had to be content with attainment as a yardstick of ability, it is now possible to crystallise the Expert's performance in the definitive terms of a memory test. It is the central theme of our examination system, and gives far better results (page 66).

ORIGINALITY

This is another of those outmoded characteristics. Nonscience prospers and proliferates because of its adherence to basic concepts, and this is obviously important. Originality is the trait of being 'new' or 'not copied' and this is clearly a contradiction to the *very fundamentals* of university training. Students *have* to copy, in lecture notes and in their memory tests (*vide supra*), the teachings of their elders. And this is how training goes on from one generation to the next. To suggest that one should 'not copy' in this way is as absurd as sending a child to a Mandarin college for him to learn Esperanto; it is like telling someone to chew and swallow a morsel of food without eating it: in short, it is self-defeating. This effete and old-fashioned 'quality' has long been overdue in its demise. Happily the adherence to timetables, notes, schedules, etc., helps to control this undesirable quality in our universities and colleges.

HETERODOXY

This comes into the same category in many ways. It is all very well for some misguided folk to talk about it as though it was to be admired, but we would do well to remember that it implies eccentricity—i.e., being *away from centre*.

Successful Experts want to be exactly *in* the centre. They do not want to be thought of as fringe people, hangers-on, someone on the edge of the action, do they?

CREATIVITY

Similar comments apply to creativity too. As an abstract 'virtue' it has nothing to offer Nonscience. Merely to 'create' something is an aimless exercise of self-indulgence—there has to be a positive, firm purpose to which any innovation can be put. Today's Expert has no time for such frivolous forays into fantasy. Instead we see an increased concentration on working from set premises, following laboratory procedures to the letter, obeying explicit instructions and generally aiming to boost the economic viability of the employer.

There is an interesting colloquial inference that shows how even ordinary Lay People think of the term 'to create', and that is in the slang usage of the verb. A mother may say of her child: 'Don't tease him or he'll only create.' It means, in this context, *to cause trouble*.

Need one say more?

INTEGRITY

And finally, what of integrity? Its medieval implications of all-out honesty are hardly relevant to today's era, where circumspection and cautious guardedness are the rule. What would be gained by announcing valuable economic advances to a competitor? Nothing at all. And what results might accrue from the insistence of emphasising drawbacks, dangers or deficiencies in any new scheme? Only this—the public would quickly leap to the defensive and progressive new ideas would quickly find themselves blocked. For today's high-speed world, in the age of the Expert, circumspection is more the rule.

We can summarise these important changes in the table.

TRADITIONAL TRAIT	CONTEMPORARY QUALITY
INTELLECT	BRILLIANT MEMORY
ORIGINALITY	CONFORMITY TO LOFTY PRINCIPLES
HETERODOXY	RESPECT FOR ORTHODOX METHODOLOGIES
INTEGRITY	CONSIDERED CIRCUMSPECTION

It is, happily for Nonscience, only a matter of time before the few vestigial vapours of respect that are still clinging to the tattered remains of the terms listed in the left-hand column dissipate altogether.

All would-be Experts must bear these basic principles in mind if they are to succeed. At all times remember that there are two fundamental aims of our Movement:

(1) THE FUNCTION OF THE EXPERT IS THE UPHOLDING OF NONSCIENCE;
(2) THE PURPOSE OF NONSCIENCE IS THE PROMULGATION OF ITSELF.

With these ideals uppermost in your mind you are already moving towards your goal.

————————— ~m~ —————————

TRAINING FOR SUCCESS, MONEY, AND POWER

Being an Expert promises money and power. That was true when the original book first came out in 1971; it still is today. The best rewards are reserved for *la crème de la crème*, who can earn over £250,000 a year. But all Experts possess power—they have unquestioned authority and remain aloof to the Lay Public. They'll say: 'Oh, I am in cancer research' or 'I work on space rockets' hoping you never find out the fact that all they do is sit at spreadsheets all day, entering somebody else's figures, and haven't a clue what's really going on.

Obedience is what matters, especially when you're doing a first degree or a doctorate. There is no point in showing creativity or originality here; you need people who will stay on that escalator and rise to the top, not dreamy innovators aimlessly charging around changing things. That's why we still have schools, even though none of them is needed now. They are unnecessary and anachronistic. Schools were set up centuries ago with breaks at Easter and Christmas, to have time to go to church and worship. They were given long summer breaks because the children were needed to help with the harvest. The idea of a class with one teacher pontificating over a couple of dozen bored youngsters is a hangover from Victorian practices. Historically, schools were the one place where you could find wise people who could explain things. They also had the right books. Nowadays, you can have the greatest teachers present you with their wisdom through YouTube. There is no need for a library, since you can consult all the information you need online. In today's world, schools have become a state-funded baby-sitting service for working parents. We have 21st-century youngsters having 20th-century people teach 19th-century methods devised for an 18th-century timetable. There is a whiteboard instead of a blackboard (that term was banned in case it was racial; though no black teacher ever thought so, and—since the blackboard was the centre of attention in a class, and the main medium of instruction—it would actually be highly complimentary if it were thought to have racial connections. And anyway, wouldn't 'white' board be just as discrimina-tory?). Teachers insist that smartphones are switched off in class. Always. In the real world, you'd teach youngsters how to use them, and incorporate them into lessons; but banning things is what school is all about.

Now you can see why it is illegal for kids to stay away from school—it is because schools rigorously instil the principles of hierarchy, conformity and obedience. People must get used to it. And you need to be obsessive. Your particular research is all that matters; it might be the tiniest sliver of the smallest subset of a minute fragment of research, but it's yours. Abandon broad-mindedness; it is the narrow focus you must master.

In the old days, schools seemed to encourage adventure. They used to have playing fields and nature tables, field trips and common rooms. Fortunately, that has all changed. Recently I was given a tour of a modern, state-of-the-art academy. It was like an office block with subservient pupils at workstations with break-out areas. To force the pupils (or 'students' as they now like to call them) into close juxtaposition with the teachers at all times, there was no staff room. I found one teacher hiding in the car park, furtively stuffing a lunchtime sandwich into his mouth, as he snatched three precious minutes alone. Everybody is given regular awards and certificates, which keeps them happy. That's so much better for the modern world. This is where people are trained how to work throughout their lives. I have often had undergraduates come to me and explain how something they're being taught is nonsense from start to finish. 'That's so out of date!' they'll complain, or 'Nobody believes that—it's rubbish!' For example, they are often taught that spelling is unimportant, so long as you can express yourself. Teachers proclaim this (though none of them would submit their child to a neurosurgeon who said they operated on the brane).

There is a survey by UNESCO on reading and writing. You can find the figures in Wikipedia, under 'List of Countries by Literacy Rate'. Nations who are embar-rassed at their poor rating (like Britain and America) have blank places in the chart, because they just don't release the figures. About a fifth of Americans cannot read or write, indeed one study by the US government suggests that half of adult Americans read so poorly that they never get the jobs they could. Look up sources like the World Fact Book and you will find the figure for literate Americans is given as 99%. Nowhere else would you find that figure (needless to say, that is an American book, written and compiled by the CIA). Currently, a quarter of all 15-year-olds in British schools have only a rudimentary ability to read and write—they cannot properly understand the General Certificate of Secondary Education (GCSE) or BTEC examinations that they have to sit at 16. There is a further great

The most literate countries in the world are places like Azerbaijan and Uzbekistan, though nations like the United Kingdom and United States pretend they're the best. A glance at the reports show that huge numbers of Westerners cannot read or write, which is a great triumph for Nonscience. This all helps Experts do whatever they want, with millions of people never realising what's going on.

example of present-day education in results just reported from Australia where many of their student teachers cannot write or count. One tutor, Melinda Wood, explains: 'Students use spell-check and stuff at home to help them, but the second they are in exam conditions they don't know how to cope.' For example, they could not calculate annual pay when they were told a weekly wage, because they didn't know how many weeks there are in a year. This problem is worsening (their pass rate fell by five percent between 2016 and 2018) but Professor John Munro of the Australian Catholic University insists that this failure isn't a reflection of the students' overall intelligence and knowledge. Quite right too! You don't need spelling to take charge. Underlings can handle all that for you.

Inexperienced students wonder whether to raise these issues and have things modernised. Bless their hearts. I patiently explain that they are obliged to learn, and regurgitate, the contents of the syllabus; nothing more, nothing less. They have to show obsessive conformity and obedience, not integrity. This is how you gain qualifications, not by trying to be honest or (worse still) clever. Raising 'issues' brings you nothing but trouble. You must obey.

One of the benefits of traditional schooling is that it stops youngsters from developing precociously in a way adults would find hard to control. You have to sit your exams at 16, and then again at 18, and you have to pass them, no matter how irrelevant they are. Many youngsters know what they want to do when they're very young. Violinists often start aged 7; artistic talent is often obvious by the time a child is 5; ice-carving masters mostly come from Paete, a town 80 miles

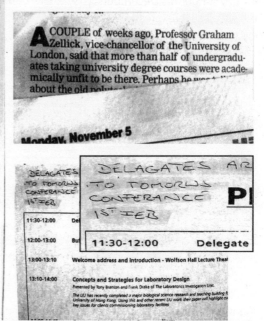

Exam boards 'told to mark down A-levels'

Geraldine Hackett
Education Correspondent

THE government's exams agency was last night at the centre of a row over claims that exam boards were put under pressure to mark down this year's A-level papers.

Independent schools suspect that large numbers of students got unexpectedly poor results because officials from the Qualifications and Curriculum Authority (QCA) intervened at a late stage to ensure that higher marks were needed for A grades.

The final results, released last month, came under fire from traditionalists because 94.3 of candidates passed and more than 20% were awarded A grades.

Independent schools now believe that the actual results were better still and that exam boards were encouraged to bring down the number of A grades at the last minute to save face.

Andrew Grant, chairman of the academic policy committee of the Headmasters and Headmistresses' Con-

summer. By making the grades harder to achieve at AS, you don't have to move the boundaries as much for the final A-level. This is unfair because some candidates will have taken AS papers in January and they would have got higher grades for the same marks," he said.

Requests for re-marks of papers have jumped dramatically this year. The largest A-level board, AQA, said appeals had tripled to 3,000.

The HMC has received scores of examples of results that were unexpectedly low. At the Royal Grammar school in Worcester, Ewan O'Farrell was awarded a C

❝ Schools believe that pressure was brought to bear on all the exam boards and across all subjects ❞

A COUPLE of weeks ago, Professor Graham Zellick, vice-chancellor of the University of London, said that more than half of undergraduates taking university degree courses were academically unfit to be there. Perhaps he ...

Monday, November 5

DELAGATES AR
TO TOMORUS
CONFERANCE
1st IEB

11:30-12:00	Del
12:00-13:00	But
13:00-13:10	Welcome address and introduction - Wolfson Hall Lecture Theat
13:10-14:00	Concepts and Strategies for Laboratory Design

Presented by Tony Branton and Frank Drake of The Laboratories Investigation Unit.

The LIU has recently completed a major biological science research and teaching building f University of Hong Kong. Using this and other recent LIU work their paper will highlight co key issues for clients commissioning laboratory facilities

11:30-12:00   Delegate

Examiners are routinely reminded to keep things simple for candidates. Only occasionally does anyone point out how this causes students to lack the abilities they need. The handwritten note on the sign for postgraduates at Cambridge University reminds us that you don't need to be literate to succeed. Every word longer than three letters is misspelt, wrongly punctuated, and poorly written.

from Manila, and they start aged 4 (how curious, when you consider there isn't any natural ice in the Philippines). Similarly, the chefs in Indian restaurants usually come from the Sylhet Division of Bangladesh, not India; and they start learning back home with their parents from the age of 3. In our world, you learn law when you go to legal school, or medicine when you become a trainee doctor. It would be much more sensible to start learning such things from the age of 7. But then, relevance has never been the keystone of training: it's conformity and obedience we need. That's how you acquire unapproachability. It would be far better for them to leave school any time from 14 or 15, if they want to launch their careers, and come back in their twenties when they want to. Indeed, the Chief Constable of West Midlands Police, Dave Thompson, said in February 2020 that forcing youngsters to stay in school, 'learning Mickey Mouse subjects' until they are 18 is a major cause of crime. Doesn't he realise how Experts work? Children in Britain have to stay in school until they're 18, to make quite sure they have had conformity drilled into them.

When you are qualified, people will react to you as though you are somehow exclusive or superior. Of course you are—but don't worry about those attitudes. People might once have teased you as being a nerd, a smarty-pants, or a geek; but not any longer. In December 2019, a psychology lecturer launched a movement to have all such expressions banned. They must be recognised as 'hate speech' so anybody using expressions like that would be breaking the law. This is just what we need!

Mail Online

Home | News | U.S. | Sport | TV&Showbiz | Australia | Femail | Health | Science | Money |
Latest Headlines | UK Election 2019 | Royal Family | Prince Andrew | News | World News | Arts | Headlines |

Calling someone a 'nerd' or 'smarty-pants' is the last taboo and should be made a hate crime says psychotherapy lecturer

- Dr Sonja Falck wants 'braniac', 'geek', 'know-it-all' all to be deemed a hate crime
- Psychology lecturer and psychotherapist says anti-IQ slurs are damaging
- Wants those with high intelligence to be given same legal status as BAME groups

By LARA KEAY FOR MAILONLINE
PUBLISHED: 01:57, 18 December 2019 | UPDATED: 07:20, 18 December 2019

Share   2.8k shares   2.1k View comments

Branding someone a 'nerd' or 'smarty-pants' should be made a hate **crime**, an academic has claimed.

Psychology lecturer and psychotherapist Dr Sonja Falck says 'divisive and humiliating' anti-IQ insults can have negative effects that last a lifetime.

She wants people with the highest IQs in the country, who make up two per cent of the population, to be protected by the same hate crime laws as ethnic, religious and sexual minority groups.

The University of East **London** lecturer and Harley Street psychotherapist is also calling for the insults 'braniac', 'know-it-all', smart-a***, 'dweeb' and 'brain box' to be covered by hate crime legislation.

Her views are based on eight years of research, which was spent speaking to dozens of high-ability children, parents, and adults about their own experiences.

© University of East London

Psychology lecturer and psychotherapist Dr Sonja Falck is calling for the insults 'braniac', 'know-it-all', smart-a***, 'dweeb' and 'brain box' to be covered by hate crime legislation

Psychology lecturer Sonja Falck made headlines recently for launching a campaign to make using descriptions like 'dweeb' or 'braniac' into hate crimes. This can only reinforce your own sense as an unassailably superior member of the human race whom nobody can ever tease. Mind you, she does look like a smart-arse. Just don't tell anybody I said so. In particular, don't tell her.

Official UK government statistics claim to report high levels of literacy. Everybody can read—well, almost everyone. The published data for 2005, 2010, 2015, and 2018 all show levels at 99% (*left*). Excellent! Spoilsports at the National Literacy Trust are ruining it all by finding out the facts: 16% are illiterate, and one-fifth can hardly read. Darn!

HIDING ILLITERACY

The problem with allowing educational standards to slip is that the results will reveal the truth. The reasoning is simple: if examination standards are so low that almost everybody passes (as we saw on page 26) then the product will be an ignorant and illiterate community, and nobody wants to admit that. The answer is simple—falsify the facts. The result is that Britain and America both claim 99%, whereas the poor educational standards in both countries show that about one-fifth of the population can't read. You thought illiterate people were rare, didn't you? Hah. Wrong again.

The United States similarly manipulates the figures. They also claim literacy at 99%, whereas the National Institute of Literacy gives the real result—19% cannot read, so roughly one-fifth are illiterate. The official figures come from the *World Factbook*, but this is published by the CIA, from which one's own conclusions may be drawn.

CHAPTER 3

How to Become an Expert

There are, by any standards, a great many Experts alive today. It is customary to say that 'of all the Experts in the field who have ever lived, 90% are living at the moment' which is very eye-catching and quite popular (especially because, like all the best pronouncements of Nonscience, it is impossible to disprove). When I was born, just as the Second World War was getting under way,* there were about 3,000 postgraduate students in British universities. By the time I was going to high school there were 8,000 of them. And now in 1970 there are over 30,000—so it is clearly a growth sector.

A report published in the summer of 1970 by London's Ministry of Technology showed that Experts were well off, too; even the lowest-paid 10% reach £2,000 a year eventually (equivalent to £40,000 in 2020). The median salary tops £2,000 (£30,000 in 2020) by the age of 35 and reaches £3,000 (£60,000) before the age of 50 according to the report.

Actually, as some pedantic people insist on pointing out, those figures are slightly suspect. Although the Ministry issued them as evidence of the re-muneration of typical individuals, it did introduce some important selective procedures into the statistical method. The survey was based on the salaries of members of the professional institutes of biology, chemistry, mathematics, metallurgy and physics; and that is a selected sample. The people who join these bodies (in some cases who can *afford* to) are those at the top end of the Expert scale; those who desire the added status and are getting on in the world. It is true that if we look at a similarly weighted set of figures to show the low end of the spectrum then salaries are far lower. Some sections *never* reached the £2,000 mark postulated by the report.

But before we denigrate that form of selection in presenting figures, let us emphasise something of the background. As we shall see, Experts are primarily

---

* There is no cause-and-effect relationship implied between these two events. However, the use of the elementary concepts of Nonscience would enable such a correlation to be established by any Experts worth their salt.

concerned with *proving their point*. Now, if we are going to prepare a publication designed to attract people into the profession, why should we consider this
lower end of the scale? Indeed *why should we even mention it*? It is safer and
more propitious by far to take the best Experts around, and work from *their*
salaries. Clearly if the Ministry had mentioned that in its report, some scat-
ter-brained publicity-seeking extrovert would probably have dragged it out into
the public spotlight as though it was a matter of controversy—and so a far safer
attitude was adopted. It is one which all Experts should learn to utilise when
there is some drawback or disadvantage inherent in a new idea:

keep quiet about it.

In countries other than Britain you can find an even better deal for the Expert. In
the United States the average salary for them in 1970 topped £4,000 (£60,000
in 2020) and in Canada it is around the £3,500 (£53,000) mark. Australia, New
Zealand and the Commonwealth countries rose just above £2,000 (£45,000)
(which was actually the average in Britain, too, but don't tell the Ministry I
said so).

So it pays to be an Expert. Not only that, but there is a great deal of prestige
inherent in the job. If you, as ordinary, boring, un-Expert Fred Snodgrass make
some suggestion to your friends in the bar, such as 'I wouldn't mind betting that
the tax bill is going to double in the next five years, hic' you would be—rightly
enough—laughed out of court. But if, as Professor F. Jaspar Snodgrass FRS
you do the same thing, the results will be entirely different. It may well get into
the newspapers then, as an Expert Pronouncement. Such statements, which
appear with regularity in the Western press, start with a set formula approved
by Nonscience practices such as—

'According to Experts, taxation is expected to …'

or

'However, the official Expert view is …'

No matter how unlikely, even outrageous, the suggestion might be, the cloak
of anonymity serves both to guard the identity of the Expert making the statement, and to give it authority. It never, simply *never*, fails, such is the natural
reverence now being paid to our Movement. A journalist who handled this
story incorrectly and gave it an unacceptable attribution would find it instantly
rejected by the editor:

'*… this startling forecast was given to me by some boozy old bastard in the
four-ale bar*' is a formula that no-one would dream of using, even though
the journalist would probably insist that it was in certain respects not wholly
incompatible with the truth.

'... *I was told this last night by a fearfully well-educated chap I happened to run into who certainly sounded as though he knew more about it than I did*' is getting nearer, but still lacks correctness.

'... *such was the claim of Mr Frederick J. Snodgrass, a reliable source of information*' simply lacks that ring of hard authenticity.

But:

'... *this is the latest Expert forecast*' or '*the latest survey shows*' is truly inspiring. It says little, gives nothing away, and is in essence no attribution at all. But it is this approach that works. It conveys the admiration for Experts that all right-thinking people naturally embody; it pays due regard to their years of training and study; and it invariably gets results.

Even on the domestic front the word is used to convey pride, humility and subservience. 'Why don't you ask our Fred?' they say. 'He's the Expert' whether it's a blown fuse that needs replacing or a row of drooping cabbage plants.

There are then, clear advantages in being an Expert. The problem is—how do you become one?

From the outset we must understand that it is not easy. The career of an Expert makes certain specific demands. It is not for the faint-hearted, the person who is quickly put off by insecure suspicions about such details as long-term risks and short-term hazards. Experts must be direct and to the point, with sufficient courage to carry their own personal ideas home no matter what.

It is not for the old-fashioned realist either, the individual who wants to complicate an issue with talk of 'social consequences', and who is easily put off by earlier published references to drawbacks in his scheme. Experts cannot afford that kind of mental laxity.

Neither can they tolerate that quaint concept of 'objectivity'.

This is an illusion, and the Expert knows it. Its practitioners (said in Nonscience terminology to be *objectionable*) are soon weighed down by their obsession to look at wider and wider aspects of the matter in hand—and that is the very antithesis of Nonscience. Experts if they are to succeed must be concerned with finer and finer detail, with matters ever more refined, and with a terminology continually more specialised than before. This is the age of the specialist (an older term for an Expert) and those who look at vast, broad issues have no place in the system.

At one time it was fashionable to question disciplines, and this will not do either. If Experts are to become the trained, conditioned people they are (an *astronaut of the world*, as we like to say) then they must learn the rules of the game: they must understand their allegiance to the Movement, their obedience to the teacher, their adherence to the principles; raising issue and doubts that

There are endless advantages in being an Expert, which have applications in fields quite apart from Nonscience. With this apparatus, our (temporarily) bespectacled friend will soon find his contact lens.

are peripheral to the main course of study—and which only make a reputation as a trouble-causer—merely sidetracks the training procedure and causes delays.

The aim of the old-fashioned 'scientist' was simply to be clever; to find things out. And that is all. It is a sterile and out-of-date attitude.

The Experts' aim, by contrast, is to progress steadily through a predetermined process of conditioning and training, collecting paper qualifications as they do so, until they reach the pinnacles of their profession. It is a secure and progressive career, with tangible—and certain—benefits. And being part of such a great Movement as Nonscience, a capital investment sector without precedent in history, Experts can feel proud that their efforts are going to be rewarded on such a vast scale. It is a brotherhood, an international fraternity with its own codes and its own mores; and a form of status that would be frightening were it not so richly deserved.

The main quality that an Expert needs is memory, the more photographic the better. Selection towards this end commences in the schools at an early age, with alphabet and multiplication tables that are committed to memorisation. Both of those are quite handy, and children soon manage to learn them.

This gives us the first selection procedure. Now come the difficult years: for as time goes by youngsters have to decide for themselves whether they are going to be practical people, always searching for a 'purpose' in what they do, or whether they have that steely tenacity of mind that allows them to memorise data and facts as a way of life. Less obviously useful tables of data are given at this stage; skull structures, atomic weights, forms of mathematical differentiation, classifications of plants and animals ... and so forth.

Some scholars now reject the training, and fall out by the wayside. The less introspective child, however, gifted with a fervour and an instinctive obedience to his or her superiors, is able to master all. To them the use of memory is a delight, and the purity of the mental process is in such a healthy contrast to the devious and uncertain ramifications of 'intellect'.

Why deal with the unknown, almost unknowable quirks of the human brain running random and unbridled like a stray mule, when it can be utilised for such crystal-clear gratifying purposes—like a well-trained sheepdog on the sunlit pastures of Scotland? We all know how the brash, left-wing revolutionary intellectual, filled with idealised notions and impracticable hopes, makes such a sorry comparison alongside the calculated, crisp efficiency of the memory-trained Expert.

But it has to come gradually. There have been efforts to introduce such subjects into primary schools—but young children in this age-group are unlikely to want to memorise data for its own sake (unless they are singularly gifted Expert material, of course). They persist in asking questions and raising pedantic doubts. So teachers are loath to include too much Nonscience at such a tender age. Indeed a new system of instruction published in 1967 showed the extent of teachers' antipathy—out of the 30,000 or so British schools at that time, only 400 had made any enquiry about the new course by May of the following year, in spite of tremendous publicity. Why? A report prepared by the Oxford Institute of Education for London's Department of Education and Science suggested that 'there are still doubts in the minds of many teachers as to the proper place' of such subjects in schools.

Quite right too.

Let us look further at the use of memory—the basic attribute of the successful Expert. You simply *must* be prepared to acquire and store the kind of information that old-fashioned intellectuals would have preferred to look up in books; it is an essential facet of one's image and a necessary adjunct to seeming properly impressive.

The clear, concise ability for recall that all Experts embody can always get you out of a tight corner in an argument, for instance. You may be cornered

in some tedious point of logical deviousness, but if you are sufficiently knowl-
edgeable you may score by the use of professional condescension.

Thus you might argue: 'Jones and MacThurber proved in 1943 that the
two factors are not invariably related' which suggests that they are in agreement with you. There are several points here. First is the use of the *negative
proposition*, a tried and tested manner of presentation for uncertain or unproven
hypotheses. Thus one can say: 'There is no evidence whatever to suggest that
after taking oral contraceptives for a considerable time, a woman's knees might
not drop off.' Or again: 'There is no evidence that the sun, at midday tomorrow
local time, will not explode,' or whatever.

The use of the two negatives in this manner does not, of course, imply anything. But it does *seem* to. And to guarantee the success of the ploy in practice,
all youngsters in schools throughout Britain are taught to memorise the rule:
TWO NEGATIVES MAKE A POSITIVE. By the time they have been conditioned into that belief (well before adulthood) they are ready to assimilate such
propositions as though they were positive assertions. The Expert must bear in
mind, of course, that they are really *nothing of the sort*.

But we digress! Jones and MacThurber, whom you are quoting in your
argument, may well have reached those conclusions—but their findings need
not necessarily have any bearing on the argument in hand. The reason for this
is simple: the ordinary Lay Person (your opponent in this discussion) won't
know anything about them at all, and so you are bound to score. You remember
the reference, and use it; your opponent cannot know about it, and will have to
confess ignorance. At once they are nailed. You have proved that, as an Expert,
you know more about such matters than ordinary people can be expected to,
and so you score a tactical advantage. They, in short, lose.*

It has been known for a considerable time that it is around the age of 11 that
a child begins to show a developed susceptibility to memorisation and a mature
ability to obey—either that or they're already beyond the pale. So the '11-plus'
examination was evolved that all children would sit when they reached that
age, to draw a preliminary line between the haves and the have-nots. Some
hard-hitting and ruthless arguments (put about, most Experts believe, by
revolutionaries of uncertain political affiliations) have succeeded in ending
this useful practice, so that selection of the elite has to be done on more subtle
lines.

Actually the new comprehensive system so much in vogue in Britain just
now has clear-cut advantages for the Expert. Subjective scholars, destined for

---

* It is worth pointing out here that Jones and MacThurber's conclusions need not necessarily
coincide with your own opinions. They might even be the exact opposite—but who is to know?
Indeed, the two writers can even be totally fictitious, and invented on the spur of the moment.
But that is a ploy of desperation, not to be accredited for normal use.

meaner tasks in life, are educated side-by-side with Experts-to-be, and in this way they are encouraged to develop reverence and respect from an early age. For them, the classes centre on practical skills; for the budding Expert increasing weight is placed on memorisation and enlightened conformity. The rise towards the pinnacles of society has begun.

The choice of this time in the development of a youngster is important. Children of 7 or 8 are still naïve enough to ask all the wrong questions—and it is only their innocence which prevents such a blatantly undisciplined mind from seeming to be, frankly, offensive. Questions such as 'Why don't cabbages walk?', 'Why don't I have feet on my arms too?' and 'How can damp grass give me a cold?' are frequently heard at this stage of development.

But by the teens, all that has changed. There is now an awakening of the mind to the need for conformity. The same child who used to regard girls as 'cissy' and soccer as 'daft' will now spend many hours boasting of his proficiency on a park bench, and memorising the names of victorious members of leading soccer teams. One would be quite justified in pointing out that these subjects, at that age, have an essentially arbitrary significance. But we must understand that this mode of behaviour is not basically motivated by sexual awareness or by an inborn sporting instinct.

On the contrary, it is (like most other passionate beliefs that appear at this time) born of a desire to follow accepted patterns of behaviour. It is because of a tacit acceptance of 'fashionable' conduct standards set by older children. Children in this age-group develop sudden and apparently inexplicable attachments for objects which serve as symbols of conformity to the group belief—partisan feelings towards pop stars, poets, revolutionary figures in politics, drugs and so on. This will not be because of inherent merits in the fixation objects themselves, but is due to the accepted codes of behaviour that prevail.

This is the dawning of a fundamental concept of importance throughout the entire Nonscience Movement—we know it for short as *Quasi-notional Fashionistic Normativity* (often abbreviated to '*Fashionism*', but more correctly designated as the *QuFN factor*).

As a result of this proclivity, young Experts may be quickly identified. They are—with Fashionism now implanted in their minds—willing to undertake tasks that may seem to be pointless at the time. They will memorise long tables of data that are not of immediate practical significance. They can be made to regurgitate accounts of experiments they do not, as yet, understand. They will assimilate parrot-fashion dogmas reeled off by a teacher who may not understand them himself—yet there is a purpose behind all this.

For it is all aiming towards *the examinations.*

These are memory assessment exercises *par excellence*, and consist of a series of questions which have to be answered in a set period of time on a predetermined date; generally a hot summer day is chosen. It is a form of initiation ceremony, a kind of obstacle race. Those who are faint-hearted, and suffer from 'exam nerves', are doomed to failure—and it is just as well, too. They are just the kind of people who chicken out at some crucial moment in a leading experiment. Nonscience is not for the nervy!

Those who don't agree with examinations are eliminated here too. I suppose in a way one must admire, to a slight extent, their courage in willingly walking off the promotional ladder like that; but they are obviously destined to be ill-disciplined, feckless individuals and clearly they deserve to fall by the wayside too.

We also exclude the reformist element, who allude to 'improvements' in their written answers. Experts are members of a trained and highly selected group—and they know better than to make that kind of empty criticism. And of course we also leave behind those who simply do not have a good enough memory. You're going to need one of those later, so it's as well to find out your limitations before it is too late.

We end up with the elite, the cream, the upper crust.

It is important to recognise this purpose in examinations. Many people (mostly the kind of anti-Expert we could well do without) have tried to decry them in recent years but in fact examinations are vital in many fields. In industry many things are examined, after all. Boxes of chocolates, pots of meat paste and packets of condoms alike are examined, and we are glad they are. *Devoutly* glad, in some cases. Qualified scholars are an end-product too, and it is only fitting that they should be examined at the end of the production line.*

The problem is to select the right form of examining technique, so that the right characteristics are chosen and identified.

If a pot of jam comes from a factory where 'stringent tests' are carried out, one would expect that they would include the inspection of technical processing to ensure perfection, colour, taste, fruit and sugar content, lack of toxic residues etc.—tests that are relevant to the product, that is to say.

One would hardly accept as 'examined' a pot of strawberry preserve that had been checked for octane rating, sewer sludge residue, streamlined shape and resistance to the effects of super-saturated potassium thiosulphate gel—tests which apply to entirely different products. And it would be patently absurd to examine boxes of chocolate for water-tightness, pots of meat paste for alcohol content, cans of beer for lubrication or wet-wipes for flavour and then, given a favourable report, state that they had been 'passed', wouldn't it?

---

* Batch-sampling is used for many products; for an account of this principle applied to university examinations see page 71.

Examinations hang over the head of every student through the whole of his three years. Apart from finals, which usually test him on the entire work of his course in a few hours, he will have other exams every term. If he does badly at any of these he may be thrown out of the university.

Many students complain that some exams have no relation to their studies during the year, and seem to be designed simply to catch you out. Others seem to be set only to check that students have been to all the lectures, and take no account of original thought and wide reading.

Certainly, many call for a series of facts, memorised by heart, to be written out by the student. And instead of being a stimulus to hard work they encourage students to memorise a vast range of undigested facts the night before the exam.

doctors commonly wait five, 10 or even 20 years, during which time they must watch their tongue, do their work, pass exams and generally convince their elders that they are suitable candidates for glory. They are poorly paid and often exploited, but when, aged 35 or 40, they cross the line, they promptly forget about injustices.

RIO DE JANEIRO.—About 15,000 students followed by nuns, priests and other sympathizers, have marched through Rio de Janeiro in a big demonstration called to support the students' demands for a new deal in education

EXAMITIS: Disturbed sleep, stomach pains, loss of appetite, vomiting, loss of confidence and sleep walking are some of the effects on children of the 11-plus examination, according to a survey by two Belfast doctors.

Sixty-five second-year economics students at the City of London College plan to boycott internal examinations next week.

They list five reasons for the boycott : —

1. The forthcoming examinations are considered a waste of time ;

2. They interrupt a natural rhythm of study ;

3. They are no basis for assessing the potential of students.

SCHOOL-LEAVERS were being treated in a "quite uncivilised " way under the present education system, the director of education for Merthyr, Mr. D. Andrew Davies, said in Cardiff on Saturday.

How the press regarded examinations in the 1960s.

This is just what lies behind examinations in Nonscience. History scholars have to be content with learning about history, and artists with learning about art. But potential Experts are being selected for stamina, memory and obedience; and the subtlety of their carefully planned examination programme may well become apparent to them only after they have graduated.

How many young biologists have wondered why one of their essential qualities must be the ability to draw? What chemists have felt that following recipe-book instructions is irrelevant to their studies? And how often have

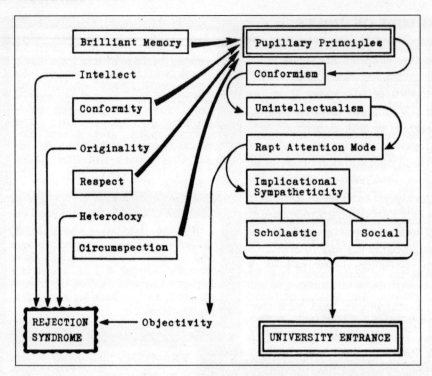

The flow diagram that governs selection at this early stage of training and indoctrination. Note the loopholes that are indicated. Those who fall into the old trap of trying to be clever, of using unbridled 'originality' as a means of impressing gullible Lay Persons, are as surely doomed to be rejected as is a transplanted heart. It may take time, but doom is certain; so avoid these pitfalls at all costs.

physics scholars wanted to break out of the rigid limits enforced by the syllabus, to stop memorising data they may never use, and to speculate idly instead? Now you can see the reason behind it all.

Those who wish to deviate may do so, of course; not everyone is fit to become an Expert. But they do so at their own risk and are *bound to lose in the end*.

We will see more of the qualities of Nonscience actually in action later in this survey. At least, you will now be able to perceive the essential nature of the discipline and the fundamental principles that underlie the selection procedures. It is a modern and progressive Movement, and one which is already ruling the entire world.

And its members—the Experts—are packed with a thoroughness and a devoted zeal rarely found in other fields of endeavour.

THE LURE OF UNIVERSITY

Why do youngsters go to university in 2020? Everyone thinks it is to acquire academic skills and concentrate on learning. The truth is very different. Teenagers become university students for clear-cut reasons: they want to get drunk, get high, get laid, and play computer games all night. It's better than getting a job. I once listed these reasons in an after-dinner speech to Cambridge University students. They cheered and gave a standing ovation, while their crusty old lecturers looked embarrassed.

These days you don't have to attend lectures, and, no matter how badly you do in coursework, even if you're still semi-literate by the end of it all, you're almost certain to get a degree of some kind. It's expensive, sure; but the money is handed out by a loan company to youngsters who know nothing about handling finance and, if the kids end up employed in a dead-end job with low pay, then they won't have to repay the money anyway. Enrolling at university postpones the need to face the real world by three years. Brilliant.

In America, you can get into university purely on the basis of athletic performance. We are familiar with happy numbskulls who were admitted because they were good at football (not 'football', *football*) or rounders (whatever they call it now). Undergraduates are being admitted for being good at computer games! It is true: The University of California now has an arena at Irvine for enthusiastic gamers, while the University of Utah runs a large sports school and offers scholarships for video gaming. The first in America to offer gaming awards was Robert Morris University in Illinois, where kids who sat up all night on video games can claim $19,000 a year for their skill. What a wonderful world this is!

It may seem curious that you can have your offspring accepted by a university in the United States because they are good at football or computer games, but even this isn't obligatory. It is perfectly possible for youngsters to gain acceptance at an American university even if they aren't capable of doing anything at all. This is done through a system of bribery, or Constructive Charitable Subvention as we call it, and centres on giving large wads of money to the university. It usually involves the payment of an administration fee to the individual who sets up the deal. That's only fair. For instance, in 2019 a football coach earned a $400,000 honorarium for facilitating the payment of $1,200,000 to Yale that would guarantee the acceptance of a student who would never normally make the grade. Annoyingly, the FBI found out about it. They intervened when a further $450,000 payment was due to be paid in respect of another useless student whose parents wanted to get them into Yale. Some well-known actors were said to have tried to take advantage of these schemes. They include Felicity Huffman, who featured in Desperate Housewives, and Lori Loughlin, who appeared in a television series called Full House. With her husband, Mossimo Giannulli, Loughlin reportedly agreed to pay a bribe of $500,000 in exchange for having their two daughters qualify for entrance (even though, in the normal way, they didn't). They were caught and prosecuted.

Many well-respected American universities are using their powerful resources to operate highly enterprising money-laundering operations. Currently, the United States Department of Education calculates that the astonishing total of some $6,500,000,000 (six-and-a-half billion dollars) has been squirreled away by universities which have received the money illicitly from overseas sources, but have tried to hide the fact. But somehow, outsiders got wind of the situation and the Federal authorities have started sniffing around. According to a report in the *Wall Street Journal*, the authorities are accusing these great universities of begging for illegal money from foreign governments and overseas companies. Many of those are known to be hostile to the United States and are anxious to steal research secrets and spread propaganda favourable to their interests. Don't think this is exclusively an American activity. In February this year Jesus College, part of Cambridge University, published a 'white paper' in support of the Chinese telecoms company Huawei. That was something of a surprise, until it was discovered that they had been secretly handed £155,000 by Huawei and another £200,000 from the Chinese state. Journalists are trying to sniff out details like these.

This is going to compromise even the top universities. It is hard to know which university is the best in the world. All the American league tables put their own universities at the top, whereas the rest of the world knows that the best are English (just as America has a 'World' Series for baseball, in which the competing teams are American). Harvard is usually top of the American universities, which you might think was wealthy enough to be above money laundering, but they are being investigated for concealing illegal money handed over by the governments of Russia, China, Saudi Arabia and even Iran. The authorities seem to think that Harvard has been doing secret deals with the Chinese telecom giants like Huawei, with Russian organizations including the Skolkovo Foundation, and even the Alavi Foundation that has links to the government of Iran. Questions are being raised about Stanford, and Yale is now being asked about huge sums of money secretly obtained from countries like Communist China and Saudi Arabia. Between 2014 and 2017 Yale decided not to submit any financial disclosure forms, and the authorities now think that more than $375,000,000 in illicit funding has been squirreled away. This is a worrying trend, Experts; not the hiding of vast sums of money, but the fact that the Federal authorities are investigating. Please be more careful. Financial manoeuvring is fine—but you must always ensure that you are never caught out.

We can find related examples which exemplify the idea. For instance, there was a tremendous outcry when an English football team of whom you may have read, called Manchester City, was found to have received more sponsorship money than it was allowed to do under the regulations laid down by the Union of European Football Associations (UEFA). Manchester City is owned by a company based in Abu Dhabi, founded by Mansour bin Zayed bin Sultan bin Zayed bin Khalifa Al Nahya, who is not himself English. Generous personal payments are often made

throughout the Arab world to encourage success, and to help burgeoning organisations achieve their aims, though matters are viewed differently in Europe where
we call such things 'bribery'. If UEFA hears that its financial rules may have been
broken, they are likely to investigate. And this is what happened. The Adjudicatory
Chamber of the Club Financial Control Body (I am not sure who they are, and I
am getting lost here) said the club had been 'overstating its sponsorship revenue
in its accounts and in the break-even information submitted to UEFA between
2012 and 2016', whatever that means, and that they had 'failed to cooperate in
the investigation'. The immediate decision of UEFA was to ban the team from
taking part in the following two years of their football matches, and Manchester
City were also fined £25,000,000. Now, the money they had been given by their
Arab owners recognised achievement. It was there to support the club in making
further progress and, let's face it, it was given only because the players were
playing football extremely well. It wasn't in any way a bribe, nor was it an induce-
ment; it was a reward. People like rewards. Yet the club suffered enormously once
they were found out, even though what they have done isn't remotely as devious
as what those universities do all the time. So—why did they suffer so egregiously?
The reason is a lesson to us all: they weren't benefitting Nonscience, and they
were not producing Experts. We cannot have people taking advantage of financial
impropriety in the world of sport, because sport offers role models to the public
and entertains the masses. It stops them worrying about more worldly matters.
The world of the Expert is above all that, because they are into power, control and
personal advancement, not mere entertainment. Bear that in mind—and, whatever you do, don't get caught out.

The class of degree you can obtain at a British university is now changing
rapidly. The norm is now to give everybody an honours degree, no matter how
badly they've performed. Once, a first-class degree was reserved for the top of the
heap; now the proportion awarded each year is steadily rising. It isn't a sneaky,
subtle increase, but a massive onward march. When *Nonscience* was published,
only 7% of students gained a first. Over the next 40 years that doubled to 14.4%.
It took less than 10 years for that number to double again, and it is now 28%. Two
universities are heading towards giving every student a first-class honours degree:
the University of West London (yes, there is one) and Wales Trinity St David (no, I'd
never heard of it either) are on schedule to give every student a first-class honours
degree within 10 years. Imagine! Everybody will have the same kind of qualification and they'll all be, what, £100,000 in debt by then? Obedient graduates,
trained and subservient, and all deeply indebted to the banks before they start
their careers. That's the way to do it.

In 1970, When the original *Nonscience* book was being compiled, it seemed
remarkable that the number of undergraduate students in Britain had increased
from 3,000 (in 1940) to 30,000 (in 1970). The mushrooming growth predicted in
that book has come to pass ... in 2020 there are 600,000 of them! Altogether,

there are currently 2,300,000 students in higher education in British universities. When I used to visit communist countries, back in the 1960s, they used to boast about an aim to have half their population go to university. How ridiculous that seemed. Yet we were soon to follow. Some 20 years ago, the labour Prime Minister Tony Blair said he aspired to seeing 50% of all young people in university. In September 2019 figures were released saying that astonishing target had been exceeded—the proportion was then 50.2%.

I have no idea what proportion of the general population are potentially academic. It might be 2%. It could be 5%. At a pinch, approaching 10%? I am certain, though, as are you, that it's nothing like half the population. Experts and academics are respected far more than people like bricklayers, plumbers, electricians, even though these are the people who keep society running. It is important that vocational activities are always regarded as inferior; we don't want the public to get above themselves, do we?

The traditional purpose of a university was to educate academics and, for as long as anyone could remember, a university education had always been free in Britain. That made perfect sense: graduates tended to earn more, and contributed far more in taxation, and so they paid handsomely for their education. But change

Sir Winston Churchill was one of the greatest minds of the twentieth century and became renowned for his favourite spare-time occupation—building walls. Although historians are too polite to say so, the results were crude and unprofessional, reminding us that the skill of our artisans often transcends that of our leaders, no matter how clever they are. Never mention this to anyone.

was afoot. In September 1998 the socialist Prime Minister Tony Blair introduced the astonishing idea of British students paying to be taught, with each undergraduate contributing £1,000 at the time. That steadily increased, until by 2015 the fees had topped £9,000 per year for tuition alone. Don't think this means the students are being better taught! Lecturers hardly know their students these days, and they certainly don't give time like they used to in the old days. One young student studying for a BSc tells me she just asked if she could have a chat with her lecturer about a few points that weren't clear; she was told there was a limit on such a consultation of 10 minutes because 'it wouldn't be fair on the others.' She is paying the college over £9,000 each year for the privilege of not being able to consult her lecturers. Most students accept what they're given but every now and again some young trouble-causer takes action against their university for the poor quality of teaching, or sexual harassment, whatever. In this case, you should buy them off—but always ensure they are prohibited from telling anyone. In February 2020 the BBC gave the game away, by reporting that one-third of British universities had been forced to pay compensation to grumbling students since 2016. The total amount of bribe money paid out was £1,300,000 with the highest sum being £40,000! Student complaints can cost the university—but nobody is supposed to know it happens. It's students who are meant to pay up, not their colleges.

Currently, the typical graduate leaves an English university owing £50,000 to the banks with some hapless youngsters eventually owing £100,000.* The money was originally handed over as student loans with interest at 0%; now interest charged is over 6%. This system has several advantages. First, it ensures that the universities are given huge sums of money borrowed on their behalf by impoverished teenagers (the highest-paid university chief in England scoops over £800,000 in a single year). Second, it gets students accustomed to being in debt from the start. Third, it has them used to paying interest. For modern capitalism to work, the population needs to be deep in debt, and people must be paying steady interest. Start as you mean to go on, eh?

Employers demand that applicants have a degree before considering them for a post. Graduates often have to take the most meagre jobs. It just shows how the public have been taken in by it all. As C. Montgomery Burns so memorably said, excell-*ent*.

THE EXPERT RISES

Once you rise above the common herd and become an Expert, you are established at the pinnacle of society. The aim (as the original book clearly taught) is that

---

* Curiously, in Britain it is only English students who pay. If you are resident in Scotland, Northern Ireland or Wales, you don't. Work that one out.

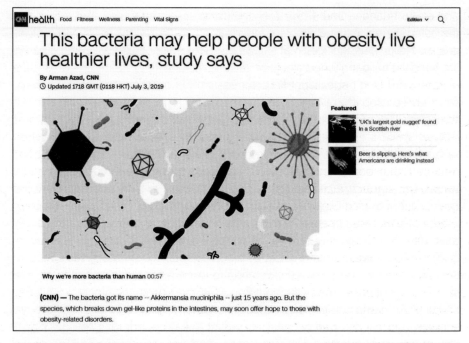

This bacteria may help people with obesity live healthier lives, study says

By Arman Azad, CNN

🕓 Updated 1718 GMT (0118 HKT) July 3, 2019

Featured

'UK's largest gold nugget' found in a Scottish river

Beer is slipping. Here's what Americans are drinking instead

Why we're more bacteria than human 00:57

(CNN) — The bacteria got its name -- Akkermansia muciniphila -- just 15 years ago. But the species, which breaks down gel-like proteins in the intestines, may soon offer hope to those with obesity-related disorders.

This recent report from CNN is an excellent example of current reportage. The heading is ungrammatical (it should be 'this bacterium') but nobody uses that anymore. Like 'the media', which began life as a plural and became a singular noun, 'a bacteria' is now accepted speech. One day 'a mice' or 'a roast geese' can be expected to appear. The name is wrong, too (it should be in italics). Even better is the image. There are no bacteria, anywhere in the world, that look remotely like these cartoons. You might as well have a picture of SpongeBob in a TV documentary and say 'this is a field mice'.

people will believe what you say. The latest 'Expert pronouncement' is believed by everyone; 'what Experts now say' guides mere members of the public in every walk of life. You must remember to use a patronising form of language, never utilised in other walks of life. Go to the doctors and they won't ask you to sit on a bench, or lie on a couch; they'll say: 'You can just pop over here.' The word 'pop' is only used when balloons burst, carbonated drinks are sold, or Experts patronise the public. People would normally ask if you had abdominal pain, but if you're a patient it will become a 'sore tummy'. You might have a bladder problem but it will be described as 'the old waterworks' and if you have a stinging sensation when passing urine you will be asked 'does your water burn?' The logical response would be: 'I don't know as I've never tried to set light to it,' but that gets you nowhere. Similarly, (and this began with the police) you aren't a *man* or a *woman*, but a 'male' or a 'female'. Using adjectives as nouns is popular with Experts, like 'the unknown' or 'the cytometric', 'the dominant' or 'an imperative'.

It is always important to use these elevated, meaningless terms. Make liberal use of phrases like: 'thinking outside of the box', with 'best practice', also 'clear and present danger' and 'fit for purpose'. Making things easily understood is very difficult; whereas making simple things seem hard is much easier. Long terms are crucial. Never have an ice-cream headache (brain freeze) when you can suffer sphenopalatine ganglioneuralgia, or feel dizzy when you can have orthostatic hypotension. Don't speak of mad cow disease when you have bovine spongiform encephalopathy available, or mention tooth-grinding when you can cite temporomandibular joint syndrome. You must make sure your words are current. Fashionism failure awaits people who use outdated terms like 'handicapped' when you should say 'disabled'. People with 'learning disabilities' used to be called 'retarded', but don't run the risk of that. Experts often use plurals as singular nouns; this started when 'the media' began to be used as a singular noun, and now you find trained Experts speaking of 'a bacteria'. I have even heard 'bacterias' used as a plural, and 'medias' is also creeping in. Sentences like 'this is an important bacteria' crop up more often than the (old-fashioned) term 'bacterium'. In fact, I can't think of anybody saying 'bacterium' for years.

Be careful how you speak. Linguists use the term code-switching for the manner in which people alter the way they address others—they change their speech to match the situation, like a child using playground slang with their friends, yet speaking normal received pronunciation English for their posh parents. Experts also code-switch, but for a different reason. They use a language of their own, full of specialist terms from ancient Latin or Greek. You are taught that such language is used for clear communication between Experts, but that's not the reason. These terms are used, not to communicate with other Experts, but to excommunicate the public. It is not a question of clarity, but preventing other people understanding anything you say.

In a haematology laboratory they spend all day working with red cells, staining red cells, observing red cells, counting red cells. But the moment someone's aunt visits, someone from management comes into the lab, or a member of the grant committee comes to call, the *red cells* become 'erythrocytes'. This means 'red cells' but says it in Greek (from ἐρυθρός, 'red', and κύτος, vessel = cell). That's one way Experts keep their work mysterious—translate anything people might understand. Translating a problem is as good as solving it. I recall a newspaper report that said scientists had discovered the nature of a disease of salmon in which ulcers of dead skin formed on their bodies. 'The disease has been identified as ulcerative dermal necrosis,' said the report (a translation of which means 'a disease in which ulcers of dead skin formed on their bodies'). If a patient comes in with a high fever you cannot explain, then tell them they have idiopathic pyrexia. This means 'a high fever you cannot explain' but says it in Greek. This leaves everyone happy. These foreign terms are to be kept within the fold and should never be used by the public. Abbreviations can be just as impressive. COVID means nothing more than 'coronavirus' but says it in code, and genes we don't understand are often known by abbreviations, like *BRCA1* and *BRCA2*. These names just mean BR (breast) and carcinoma (CA) numbers 1 and 2 (these should always be in italics). When the new coronavirus escaped from China is was soon given the name 2019-nCoV. The western media said they had now identified the virus, but the abbreviation means only 2019 (the year it appeared) n (new), and CoV (coronavirus). Nobody knew anything about it. Meanwhile, the Chinese National Health Commission said it was actually NCP, meaning Novel Coronavirus Pneumonia (which is an illness, not a virus). The name was changed again in February 2020 by the World Health Organisation, who said it would in future be called COVID-19 (CO for Coronavirus, VI for virus, D for disease, and 19 for the year). They had scores of serious experts deliberating over this for weeks. A pedant would say the name makes no sense. No other virus disease has been named according to this system, it is no better than the already accepted name, and anyway this is the name for the disease and not the causative organism. Pay no attention. Nonscience is never meant to be sensible; just to be accepted.

Coronaviruses have always been given names starting with HCoV (Human Corona Virus). And so we had HCoV-229E virus and HCoV-OC43 (both causes of the common cold), HCoV-NL63 (a cause of croup and sometimes severe chest infections), HCoV-EMC (Middle East respiratory syndrome, or MERS), and HCoV-HKU1 (another rare cause of chest infections). Then we had Severe Acute Respiratory Syndrome, SARS. It is a ridiculous name; syndromes are supposed to describe the unique characteristics of a condition, but this description covers every severe acute respiratory illness from a nasty cold to fatal pneumonia. Its name is confusing, too; it should have been HCoV-SARS, following the well-established convention for naming these viruses, but no, to keep things interesting, it was dubbed SARS-CoV, and the virus that causes the latest epidemic of COVID-19 is now called SARS-CoV-2. Yes, although Experts didn't admit the fact, the COVID-19 plague is the return of SARS, but in a form that is ten times less likely to kill those who catch it.

There was a time when viruses were named after the place they originated. We had Spanish 'flu, which killed more people in 1918 than had died in the First World War; then Asian 'flu in 1958; Hong Kong 'flu, the global epidemic of 1968; and Russian 'flu in 1977. A form of influenza which emerged in 2009 (and first flourished in the United States) would have been called American 'flu, but (since the United States likes to name these things) they insisted we called it 'swine 'flu' instead to disguise where it spread from. The virus of this latest pandemic would once have been called Wuhan 'flu or Chinese 'flu, but this is now considered racist. Nobody does that. Well, you do if you're Chinese; they call it Wuhan virus all the time.

LATIN NAMES

When it comes to confusing the public, the BBC is our ally. Much of their science (and all their biology) is wrong. One of the ways in which the BBC patronise their

Viruses were once named after the place of origin, though nobody civilised in the West calls the latest coronavirus Chinese or Wuhan virus. But they do in China. English-language posters and programmes run by Chinese people do so all the time. I particularly like the smiley face (*bottom left*).

American president Donald Trump has called the pandemic Kung 'Flu and says it is caused by the China virus. He likes people to think it's just a slip of the tongue, but sharp-eyed photographer Jabin Botsford photographed his script to show that Mr Trump had changed the word himself, just in case he forgot.

audience is a ban on Latin names for microbes. If you want to mention an alga polluting the water, then you are forbidden from telling the public which one it is. They will mention some parasite is causing problems, but you must not (under any circumstances) reveal its Latin name. Of course you cannot justify this; little children love dinosaurs with names like *Tyrannosaurus rex* or *Mamenchisaurus sinocanadorum* and even *Micropachycephalosaurus hongtuyanensis*, just as old people are happy to go to the garden centre to pick up *Metasequoia glyptostroboides* or

Hypophyllocarpodendron cucullatus. Gardening programmes on TV are the only place where producers let you use Latin names, mainly because the Latin names are the only names the plants have. But you cannot name a bacterium (sorry, a bacteria)—with one exception. You can identify *Escherichia coli* by name, but only if you shorten it to *E. coli*. This is famous as a dangerous germ. Just remember: it is typically harmless and it is the commonest bacterium in the body. One-third of faeces is pure *E. coli* and this is a beneficial microbe which helps keep your intestines healthy (the strains that make people ill are rare mutants). *E. coli* is also of industrial importance—a decade ago it contributed $500,000,000,000,000 (500 billion dollars) to the global economy. And the other important point is that it should always be printed incorrectly, usually as E-coli. That is no more correct than *president*-Trump or aA-*milne*. The name can only be correctly printed *E. coli* and just that—but these facts must be kept from the public.

Living cells are the most important concept of modern science that people need to understand, and the BBC manages to censor programmes so they hardly ever appear. Computer graphics are used to represent them, which is good because they look nothing like the real thing. The last time they included real microbes was in 2013 in a documentary series called 'Wonders of Life' hosted by Brian Cox, a physicist who naïvely believes that you can account for all living phenomena through Newtonian principles. In one episode they featured low-magnification movies of *Paramecium*, trying to show that it worked like a primitive robot that backed off if it encountered an obstacle. And where did he demonstrate the movie clips of these microbes? In a laboratory, equipped for aquatic microscopy? Perhaps in the pool of a beechwood glade where *Paramecium* abound? In fact, the production team drove 20 minutes west of Miami into the Everglades, and stopped at the tiny unincorporated settlement of Coopertown, where there is a small restaurant whose specialities include frog's legs, catfish and gator tail. They set up the presenter near a window, and projected the images of *Paramecium* on the wall, and also on his cheek, the napkin dispenser, and a bottle of Tabasco. You might have speculated that the microbes projected on the presenter's face would link to an explanation of the cell communities that comprise his head, or that the Tabasco was featured because of the microbes whose fermentation creates the sauce—but no. Cox insisted that *Paramecium* works like a simple transistor, driven by the electricity in a lightning flash. There has never been a more utterly irrelevant, misleading, or bizarre, screening of any microbe, anywhere in the history of television. But (and this is crucial) it perpetuates the superiority of the Expert, and that is all that matters.

We find 'Latin names' everywhere (they are Greek just as often, but we don't like to say so). Some are intriguing, like the fungus *Phallus impudicus* (impudent penis) which grows just like a willie in the woods, or birds of the genus *Turdus*. These are blackbirds and their allies, and the name comes from the Latin for 'thrush'.*

---

* Thrush the delightful songbird, not the annoying infection.

The BBC like to avoid any sight of microbes in their programmes, and if they have to appear then they can always be projected onto the napkin dispenser of a café and a bottle of sauce. This diverts attention from the fact that the hapless presenter has no idea what he's talking about.

Then there is *Bugeranus*, a tall and majestic crane from the Greek words *bous* (ox) and *geranus* (crane). A fungus of the family Boletaceae has been named *Spongiforma squarepantsii*, there is a ground beetle named *Agra cadabra* and a small fly called *Pieza rhea**. The German biologist Karl Wilhelm Gottlieb Leopold Fuckel has species named after him, though I won't be listing any of those. The longest name in the world is *Parastratiosphecomyia stratiosphecomyioides*, a soldier fly. Not many people have studied this species. I can see why.

COMPUTER SECRETS

In today's world, computers are everywhere. But when *Nonscience* was published, they were rare and nobody had the least idea how important they would become. I first used a transistorised computer in 1962. It was an Elliott 803 and filled a space the size of a small bedroom. It cost (in today's money) £500,000 and was programmed in Basic. People thought I was smart to be able to write operational code for it. Only later did I meet up with the man who actually designed and invented it, Iann Barron, who was only a couple of years older and far cleverer than

---

* These are all genuine names. I am not making any of this up.

When Boeing launched their Starliner in December 2019, it failed to enter orbit. Did computer control cause the problem? If so, whoever wrote the algorithms will have been to blame—but we may never know for sure. Next time there will probably be people on board, who could manually tackle the problem caused by anonymous youngsters safely on the ground.

me. Things are different now; there is 100 times as much computing power in our washing machine as we had then. Building computers is all very well; but writing software code is where the money lies. Algorithms are the thing. They are written by youngsters (who lack experience and are cheaper than wiser minds) and, if they do go wrong, nobody minds. Recently we have had several key examples of algorithms going wrong. In October 2018 a Boeing 737 Max 8 crashed into the Java Sea, 12 minutes after taking off, killing 189 people. Five months later, the same thing happened, only this time the plane hit the ground in Bishoftu, Ethiopia, and all 157 people on the plane perished. Boeing immediately blamed the pilots. These planes had been fitted with oversized engines, and software had been written to make the planes handle as if they hadn't. The algorithms compelled the planes to do what the computer instructed them to do, in spite of the pilots knowing better. It now seems that a sensor may have failed, at which point the control system software refused to allow the pilots to take corrective action, and instead the planes were programmed to crash. It seems obvious that the software should never have been written to allow this to happen. The code-writers must know who did it …

Around the same time, the *Viking Sky* cruise ship was caught in a storm off Molde, on the Norwegian coast. The storm made the ship lurch violently, when

In March 2019 the *Viking Sky* was listing heavily, and close to the rocky shore, when computer control prevented the captain from starting her engines. Everyone is trying to blame him for setting sail in bad weather, but ships are designed for that. Nobody hints at the possibility that the algorithms written to control the engines stopped the crew doing what they knew they had to do.

the *Viking Sky*'s computers detected low oil pressure and shut down the engines. Unable to keep a heading into the wind, the ship was roiling dangerously, with heavy furniture crashing about. Eventually, 479 passengers were airlifted off, many rushed to hospital. Ever since the *Titanic* disaster, all cruise ships carry plenty of lifeboats and passenger drills are held every time a ship is due to sail, but you can't launch lifeboats when the sea is rough (the crew don't like to tell you this, and passengers don't realise the fact). Had the ship been out of reach of land, the situation could have been disastrous. Here too, engine control was not in the hands of wise engineers, but of algorithms written by youngsters. Once again, everyone tried to blame the captain for setting to sea in bad weather. However, the ship was capable of enduring far worse conditions, and it may be that those algorithms were to blame, shutting off all power because the rough seas gave a faulty indication of low oil pressure. In a comparable incident, the cruise ship MSC *Opera* crashed at speed into the dockside at Venice, sirens blaring loudly as the captain was unable to stop the engines. Is there a theme here? In each case the captains were incapable of changing what the computers had decreed. And the computers were programmed with algorithms that could not be over-ridden, as if the young code-writers knew better—in all circumstances—than the wise minds that run these craft. You might think that every captain should have manual override in emergency, but this is the era of the Expert. Disasters can happen, people can die, but nobody ever knows who really set up the situation for its disastrous outcome. Truth to tell, it was probably some idealistic young code-writer who had never had any other job. Such is the world of the Expert.

This is all speculation (the tendency is always to blame the operators, and never the equipment) though there are some cases that confirm how untouchable the programmers are. Two American warships were in fatal collisions within the space of a few weeks, with the navigational crew being blamed in each case; but the ProPublica website started investigating closely, and they revealed that it was actually the computerised control systems that were clearly the cause. The first was on 17 June 2017, when the USS *Fitzgerald*, a destroyer, collided with a huge cargo vessel off the Japanese coast and nine crewmen were lost. Investigations showed that the warship was a new design, controlled by a complicated computer touchscreen that nobody understood. When the ship put to sea its operating system was way out of date, and the year before the collision there were reports that the destroyer's navigational system wasn't functioning as it should. In the old days there was a lookout (to make sure no ships were too close), a telegraph (to signal engine speed) and a helm (which turned the rudder from side to side). Now there was a screen which was covered with confusing symbols. One officer said that it was as if you sat in your car and found an iPad where the steering wheel used to be, and no pedals. The problem was that some of the icons didn't work properly, and none was fully understood by the crew. The new computerised system meant that lookouts were no longer needed and so, by the time of the impact, nobody even knew the other vessel was there.

Nine weeks later, the USS *John S. McCain* turned into the path of a gigantic oil tanker causing a collision which killed 10 more sailors. The ProPublica investigators turned up a history of reports saying: 'It's only a matter of time before a major incident occurs,' and 'Incompetence is the standard.' Crew aboard the *McCain* found that the ship didn't respond properly to the touchscreen. They would suddenly find they had no control over steering. The ship would make erratic course corrections that were so confusing that they weren't even logged. Shortly before the impact, the commanding officer ordered that the destroyer should indicate that her course was no longer under control of the crew, so the ship's lights should be changed to show 'ship not under command'. Nobody could do anything to avoid a collision. Naturally, it was the crew who were blamed, and not the teams who had installed the new control system. Even so, this was one of those unusual occasions when the person in charge did take the blame. By August 2017 it was noted that there had been four collisions of warships within a year, so Vice Admiral Joseph Aucoin, the fleet commander, was obliged to leave his post just a few weeks before his planned retirement date. Nobody seems to have taken any action against the people who wrote the code. I keep reading reports of ships crashing into each other, or ramming the dockside, with their engines refusing either to start or to stop. Is this because of dodgy algorithms controlling ships by computer? It seems that it could be (but I am probably wrong). In any event, it seems to have been incompetently written code that brought down the 2020 Iowa caucus. That really was the final nail in the coffin; the caucus has sunk without trace.

Vehicle manufacturers have traditionally hoodwinked the public into buying huge cars they do not need and which cause horrendous pollution. Anybody driving a car larger than they must is guilty of wrecking the environment, though you won't find the Extinction Rebellion protesters saying so as they drive home in massive cars they can hardly control. In 2005 the British public were told to change from petrol to diesel cars* even though diesel engines can produce far more air pollution. Manufacturers soon started writing computer code to make sure their cars passed emissions tests, even though they were flooding the atmosphere with poisonous gases that killed people in their thousands. The German manufacturers Volkswagen, BMW and Daimler led the way. As a result, their cars all passed their exhaust analysis tests with flying colours, before reverting to their normal, poisonous emissions the moment they'd left the test centre. Oh yes, algorithms can work wonders.

The aura of untouchability with which computer people surround themselves is awe-inspiring and an example to us all. God comes second best. Nobody comments on your software, because no-one understands it. Algorithms take all the decisions, without people stopping to ask—who wrote the code? Their detachment from the real world is enviable. People working on projects like Deep Mind like to say that it is their 'general Intelligence' that has created the wonderful world we see around us … without stopping to realise that the world was actually created by living things (mostly microbes). The world these people inhabit is an office full of monitors and filled with electronics. The real world, marvellous beyond anything they could grasp, is something they strenuously avoid. They're Experts!

WHEN EXPERTS HIDE

Teenagers in August 2020 had a rude awakening when their exam results were the victim of the government's latest cock-up, sorry, robust and fair reassessment by algorithm. Many lost university places! And if you want a current example of Expert domination then the pandemic is what you need. The government told everybody that their decisions were based on Expert advice. Since many Experts had different views, it was easy to select whatever the government wanted to implement for political reasons and say, well, this is how it has to be. During the early stages of the outbreak, the government's priority was to keep the economy running in the hope that it would all turn out alright in the end. There is only one course of action that will control a coronavirus: you must locate where the virus exists, and stop it spreading by quarantine. It then goes away. That's how the original outbreaks

---

* This was the idea of Sir David King when he was scientific adviser to the Tony Blair government. He assumed that motor car manufacturers were people of integrity! Imagine that. How naïve can you be? He now admits his advice was wrong and says the car manufacturers 'have blood on their hands'. Experts think he should keep ideas like that to himself.

of SARS and MERS were ended. The British government showed no inclination to order the pubs, restaurants, and schools to shut until President Macron privately informed the Prime Minister that France would close their borders to British people unless they did so. This is why, reluctantly, the government were obliged to opt for lockdown. The secret that everybody kept quiet about was that there were already 10,000 new cases every day at the time. We now know that—if lockdown had been instituted one week earlier—the number of deaths would have been halved. Although it was known from early on that old people were far more at risk than the young, nothing was done to protect them. Old people coming into hospital would be infected with the virus, and then (without being tested) they were sent back to their care homes to spread it to the healthy residents. As a result, more than half of all the deaths were among the elderly in residential homes. Worse still, their relatives were kept at bay so the old folk died frightened and abandoned by their kin. Repeated protestations produced no effect at all, but (of course) the official spokespersons knew exactly what to say. 'The government has thrown a protective ring around care homes,' said Matt Hancock, famous as the star of the amusing Hancock's Half-hour, 'from the start.' The Prime Minister echoed the sentiment and said that he had personally instituted a lockdown in care homes before it was generally forced upon the public. In the sense that care homes had few means of protection and no means of testing, the Hancock statement might seem blatantly untrue, and since no mandatory instructions had been issued to any care homes anywhere, prior to the general lockdown, the Prime Minister's statement could be seen as a misstatement too. Nonsense: it simply reassured the public and reminded them all that governments know best.

People have to realise that it is up to Experts to control what goes on, and to hide behind a cloak of secrecy when matters get out of hand. That's how the system works. It is the bedrock of democracy.

Barcode: {_dlc_BarcodeValue}
DocId: 2QFN7KK647Q6-519654008-245676
Government Body: BEIS
Security Classification: OFFICIAL

Scientific experts: Patrick Vallance (GCSA), Chris Whitty (CMO), Jenny Harries (dCMO), Angela McLean (CSA MoD), Charlotte Watts (CSA DfID), Osama Rahman (CSA DfE), Graham Medley (LSHTM), Julia Gog (Cambridge), Neil Ferguson (Imperial), John Edmunds (LSHTM), Peter Horby (Oxford), Ian Young (CMO Northern Ireland), Rob Orford (Health CSA Wales), Sheila Rowan (CSA Scotland), Nicola Steedman (dCMO Scotland), Jeremy Farrar (Wellcome), Michael Parker (Oxford), Ian Boyd (St Andrews), Russell Viner (UCL), Chris Bonell (LSHTM), Rebecca Allen (Oxford)

HMG & DA Observers: Vanessa MacDougall (HMT), Lorna Howarth (DfE), Ben Warner (No. 10), Dominic Cummings (No.10), Jim McMenamin (Health Protection Scotland), Jonathan Slater (Perm Sec DfE)

SAGE secretariat: Simon Whitfield, Stuart Wainwright

(*Left*) Pressure from the press forced the government to release some of the secret documents that had been circulating around their departmental committees. We don't like doing this, since transparency is the enemy of Experts everywhere, but compromised by blacking out the text. Many people didn't like this, but what do they know? ('Not much' is the correct answer). (*Right*) After weeks of protesting, the press were able to force the government to disclose the names of the people behind whose advice they were hiding. But still we had the last laugh—many of the people, some being close friends of the author, refused to admit they were involved and so the documents were still redacted.

CHAPTER 4

The Training Process

The coming of Nonscience has been a recent event. There have been scattered Experts throughout the past few centuries (page 200), but it is only since the beginning of the 1900s that the number has become significantly larger. Very considerable numbers of Experts arose during the war, as a result of the stimulus inherent in the situation. This tradition has been carried on in the Western world ever since, and the employment of Experts by military authorities continues to be—particularly in the United States—a leading source of public funds diverted into the coffers of the Nonscience Movement.

Yet in the 1960s a swing away from such subjects began in the schools and colleges. Many of the young have simply not the stamina it takes, true; but some have even suggested that there is something 'not nice' about our Movement. Happily their pessimism and presumptuous impertinence is self-defeating in the long run, for the adherents to Nonscience have seen it advancing by leaps and bounds—so one need not fear for the future on that score.

But it also confers a considerable advantage on the would-be Expert. There are plenty of places in schools and colleges available for all as a result. Competition is slight, and chances of admission are now better than ever before. Partly this is due to the unprecedented publicity that the subject has attracted. It has certainly become Fashionistic in the 1960s—indeed it was adopted as a leading platform of campaign by the British government—and unprecedentedly large sums of money have been made available as a consequence.

The statistics that are available show the extent of the 'swing'. Since 1960 the proportion of school-leavers who were specialising in subjects such as biology, chemistry, physics and to an extent mathematics too, has declined relative to other subjects.

From 1962 to 1967 the proportion of British scholars starting A-level courses went up from 11.5% to 16.5%; pupils specialising in the arts, literature, classics and languages climbing from 6% to 9% whilst those in our field stuck at around the 5% mark. Clearly they did not have the sense to recognise a good thing when they saw it.

And what of the teachers? A quarter of all teachers in this category will be retired by 1978, and the standard of degrees possessed by teachers of the time has been steadily dropping. The total secondary school population during the period 1968–1978 has been calculated to increase by 50%—resulting in a shortage of teachers.

Those who try to suggest that this is because of 'disillusionment' with the subject, or a 'fear' of its consequences—even a belief that it is in some way 'inhumane'—are merely trouble-causing rabble. Nonscience will evolve its own way of dealing with their ilk ere long.

We have already seen the general theme behind the training of young Experts. Yet one important practical factor remains: how do we assert our allegiance in a form that the teacher will recognise? There are several vitally important modes of behaviour that are currently in vogue. They are generally referred to as:

THE PUPILLARY PRINCIPLES*

(a) Conformism

The pupil must readily and willingly conform to what the teacher says. A failure to do this will inevitably result in the initiation of a Rejection Syndrome,† with consequent risks of academic failure.

(b) Unintellectualism

Do not on any account attempt to dabble in aimless, outdated 'intellectual' discussions. These reveal an undesirable element of ill-discipline, and introduce an element into the lesson which teachers will find quite beyond their grasp. Why antagonise them with something best discarded, along with the other failings of innocent childhood? We do not suck our thumbs in Nonscience!

(c) Rapt Attention Mode

Hold yourself at all times with a briskness and bearing calculated to reveal a configuration of alert respect, awe, and reverence for the Nonscience Movement. The sitting position is all-important: those dubbed the Curate's Crouch and the Scrubbers' Sprawl are amongst alternatives that are to be avoided *at all costs*. This is the first lesson in the cultivation of the correct image: it is the first decisive, self-controlled move towards Expertism at its best.

---

* Not to be confused with Pupillary Principles in Ophthalmics, of course.
† This has no connection with the Rejection Syndrome of Transplant Surgery, *q.v.*

(d) Implicational Sympatheticity

A cultivated sympathy with the teacher is all-important. It need not be emotionally motivated (i.e., real) but it is exactly this attribute which augurs well for a close teacher–pupil bond and that in turn is essential for a good academic report at the end of your period together. It is also a necessary adjunct to successful Expertism, too; and so at the school level—apart from the official ostentation of examinations on the syllabus—we have a splendid opportunity to find out what the world of the Expert is like, and how best to utilise its potentialities.

The nature of Implicational Sympatheticity is two-fold at this level in school: (1) it concerns beliefs in particular scientific or moral issues of a SCHOLASTIC nature—e.g., the use of atomic weapons, eugenics in human affairs, relationships between the Expert and society; and (2) (on a different plane) it is also interpreted in relation to the teacher's personal life, i.e., at a SOCIAL level. Examples include: why teachers cannot enjoy protracted holidays, the reasons why teachers as a whole want better pay and working conditions, and so on.

ABOVE ALL: do not waste time arguing about what you are taught; many young people try to do this when (as is inevitable at this still early stage in training) they do not understand the rationale of it all. Leave questions (one hesitates even to suggest the word 'doubts') about examinations and the syllabus to those better experienced and more trained to answer.

ALWAYS: learn what you are told to learn, believe what you are led to believe by the teacher; and it is imperative to stick to the syllabus. Going beyond it (as a result of 'broad reading') does not qualify you for any extra marks, since the papers are marked on the syllabus only. Not only that, but the examiners will not know anything about what you are describing and they will therefore, like any trained Expert is bound to do, assume you are *wrong*.

Obedience to the above principles will, on the other hand, ensure your success. So let us assume that you have passed through the harrowing experience of intimate contact with uninitiated schoolfellows, destined for lesser things, and have emerged at the end of the scholastic phase with the correct examination passes.

You can now feel your destiny a little: the Expert is beginning to emerge. A good many Negativistic Proclivities have been permanently eliminated. Above all, the potential student has shown—and indeed demonstrated—a willingness to become totally subservient to the ideals of the Movement.

The student then enters university.

At once the significance of the well-coordinated Nonscience community becomes apparent, for individuals enter college in the first place only after filling in application forms and attending interviews. They have, of course, been taught nothing about either. They do not know the difference between a reference and a testimonial. They have no notion of what a 'session' is, or the difference between second-year finals and getting a two-two; they have never been taught what a 'fresher' might be and they do not know the first thing about the Union. However, they have (*and this is important*) been taught to expect things to be hopelessly confusing, and so they are not greatly perturbed.

All is gradually brought home through a series of rules and regulations which make the essential code of behaviour quite plain. And from then on students (as they now are) can proceed confidently, along predestined pathways of crystal clarity.

Essentially the patterns of educational technique in the university are similar to those seen at school though they do, naturally, progress systematically.*

The underlying principle is brought out in the 1968 *Enquiry into the Flow of Candidates in Science and Technology into Higher Education*—the Dainton Report—which includes the words:

> *It is most important not to equate intellectual rigour with excessive reliance upon long periods of routine experiment, upon reiterated formal exercises based on elementary theory, and upon the committing to memory of large quantities of factual information which can readily be derived from basic principles.*

Indeed, indeed; it is *vital* not to confuse the two! 'Intellectual rigour' is—what? Merely a phrase to describe the former mental attitude that we knew as 'science'. Nothing more than mental athletics of a peculiarly self-gratifying nature, with little to offer to itself or to others.

It is, in the words that the good professor coins to describe them, 'long periods of routine experiment', 'reiterated ... elementary theory' and 'memory of ... basic principles' that are the very cornerstones of Nonscience. Yet as that report implied, the two approaches are still confused by some—which is most surprising, particularly when we reflect upon the deep and fundamental diametric differences that separate them.

Even Dainton confuses them in some instances, as when he uses the phrase 'imaginative controlled curiosity'. Curiosity is not that which can be controlled. It is 'imagination' and 'curiosity' that belong to science; to the history books. It is these very qualities that make us realise one of the truths of our Movement:

Science is an art.

But 'control'—self-control, budget-control, community-control and the rest—is diagnostic of Nonscience.

---

* Often they progress unsystematically too, now I come to think of it.

Our analysis of the birth of Nonscience explains, in a single telling moment of truth, why such a confusing juxtaposition of terms has been used in the past. We must resist with all our might the peculation of our own principles by disciplines less worthy.

The exact form of control exerted over the Nonscience trainee, the student of Expertism, is not the kind of precise regimentation that one can dissect and quantify constructively. That is obvious. If one could, it would clearly not be Nonscience at all. It has not—like so many movements—been decided upon by some trivial working party or established on the basis of old-hat traditions. Rather it has arisen spontaneously, as the result of pragmatic need. It has evolved according to the dictates of circumstance, the laws of self-preservation, the Darwinian (now *there's* an Expert for you!) concept of survival in a progressive world.

Thus students at university are shown how to be calculatedly long-winded at one moment, yet clipped and pedantic at the next, dependent upon the dictates of the circumstances. They are taught how to use particularly sophisticated grammatical constructions and what appear to the Lay Person as hopelessly confusing terminological conventions. They are, as the student is taught, merely part and parcel of their vocation; a necessary means of avoiding duplicative repetitional restatementalised reiteration, and—secondarily—a means of marking them out from their fellows. The terms of the rugger player, the darts enthusiast, the skier or the pigeon-fancier are all esoteric in their own way. It is only to be expected, therefore, that not everyone can at once follow the Expert.

So much for the subconscious training; let us now turn to consciously motivated aspects.

The most important single quality which students will need—which they require above all else for hours, even weeks or years, on end—is one which they have already begun to take in hand. And that, you recall, is their memory. It has naught to do with the *understanding* of subjects of course, and therein lies a popular misconception. Experts aren't *under* anything. They are superior to all. And they will not be wasting time in *standing* when they can steadily march forward to richer rewards! So the notion of 'understanding' is naturally beneath their dignity. Anyway, the merest attempt to systematise in detail the machinations of Nonscience would soon lead to an inevitable cerebral constipationary response, since the processes and products of our Movement are so diverse, and so essentially random, that any attempt to grasp their significance in systematic terms is doomed to failure from the start. They have none.

This inborn tendency to negative feedback, towards Randomisation, is at the hub of the Expert's activities.

As Darwin showed* it is only when we have a varied stock that selection can occur, and thus the 'genetics' of Nonscience would soon lead to a weakening without Randomisation. There are many ways in which this important element is introduced, ranging from the healthy heterodoxy of the syllabus to the arbitrary free-and-easy nature of student selection.

The main categories of activity underlying the compilation of the university training scheme are as follows.

DATA SELECTION

This is the process of deciding what material should be committed to memory. It is a largely randomised procedure. No attempt should be made to 'rationalise' the matter in terms of benefits to the student, the imaginary benefits to society in general, or the best interests of the subject itself. Many courses have, in fact, derived from the dictates of mediaeval colleges in a self-sustaining evolutionary pattern. There are thus many delightful idiosyncrasies.

The kind of material that has to be learnt is not easy to list since it covers a vast area (albeit patchily) and no selected list could hope to be representative, let alone fully comprehensive. However, the following examples—chosen in a correctly randomisated fashion—will give a rough idea of what you are in for:

- tables of classification groups of tropical plants;
- outline drawings of skull structures, with particular attention being paid to the correct way of drawing suture lines as little wavy traces, and dealing in detail with fossil reptiles, etc. (though the human skull is dismissed briefly);
- atomic weights and numbers (lists of which are found in all routine textbooks);
- the histology of fossil plants from Carboniferous strata;
- detailed stages in the decay of rare heavy metals;
- refining methods used to extract elemental samples from ores; and so forth.

Many young nurses in 1960s Britain learnt the principles of leeching during their training, which shows how traditions have been preserved by the Movement. I suppose that, if Fashionism took another twist, the bleeding of

---

* Charles was one of our pioneer Experts. It is quite true, of course, that a great many men had worked on the theory of 'survival of the fittest' before him and that, in a conceptual sense, his work was entirely unoriginal. But only he realised the timely degree to which Fashionism of the subject was at its height when he published, and only he managed to present the data in such an eye-catching and interesting form. 'Survival of the fittest' was then as catchy as 'tiger in your tank' was for the 1960s.

patients with leeches could even be revived, in which case the training would be doubly valuable.

Subjects not found in the selected data include the use of libraries, types of journals available, compilation of bibliographies, how to carry out research, presentation of papers personally and for publication, protocol within the Establishment, etc.; but with randomisation—and with the time available—these things simply have to be left out. It all comes in its own good time.

These examples illustrate several vitally important qualities that are invariably found in the field. The selected data should, in general, conform to the following criteria. They should be:

(1) the kind of material readily available in any standard textbook, in order to assist accessibility during the arduous months of memorisation that lie ahead;
(2) repetitious and boring in nature—i.e., the sort of data that would only be assimilated by a truly loyal devotee of Nonscience; all others would quickly lose heart and fall by the way; and
(3) abstruse and rarely utilised in practise, thus making the information of special rarity value.

Thus we may see at once the three important classes of material included: accessible coherent, and sophisticated.*

But so much for the selection of the material itself; let us now turn to the means of presentation.

INDOCTRINATIONALISTIC METHODOLOGICAL MODALITY

This—known in 'Nonscience shorthand' as Indoctrinational Methodology—is the means of imparting information to the student. It has been developing over a great many years, though it is only in the past few decades that the most refined forms have become established.

Partly this must be because the truly significant propensities of advanced technological research have manifested themselves in a tangibly recognisable conceptual format (or have *developed*, as the uninitiated might say). There was little point in giving humankind the means to consider widespread alterations to their environment, for instance, before the apparatus to do it was available.

But now the range of available equipment is formidable. Alteration of the entire atmosphere, raising of H_2O vapour and CO_2 levels, for instance, or

---

* Sometimes dubbed by our enemies 'superfluous, repetitious and irrelevant'. How *little* they understand ...

the liberation of heavy-metallic vapours such as mercury, is one example. We can now induce diseases through entirely artificial means, bring about genetic manipulation on a hitherto undreamed-of scale—in short, we may now look forward to the day when we can irrevocably alter our planet, when visitors from afar will see the marks of Nonscience on the face of the globe, and indeed on that of the humans that inhabit it. And so it is at this time that the potentialities of Nonscience and the training of the Expert have assumed peerless importance.

Indoctrinational Methodology—In.M.M., for short—is founded on the very best of modern training principles. We must at the outset contrast this with the older concept of *education*. That term derives from the Latin *educare*, which means 'to draw out'. This spineless and wishy-washy concept had its day a century or so ago, when that form of gentility could be tolerated. But now that the need for educational conformity is at its height, and our whole concept of the tenor of 'education' has salutarily changed too, a quite different terminological expression is needed.

And so we have adopted *training* in its stead. This is a far more vigorous, healthy and outgoing expression, deriving from the Norman French verb *trahiner* (to drag). That is a far more Positivistically Orientated Conceptualisation (PoC, as we like to call it) and better fits its Nonscience context.

Information in the training course is imparted to students in several different ways, all aimed at the infusion of their very soul with the purity of the essential properties of Nonscience.

First, there is the textbook. This contains all the data they need to learn and could, from a merely practical viewpoint, be the only source of information the student requires—but of course there's more to Nonscience than that!

These books are written by Experts, who are otherwise pretty well unoccupied at the time and, since they are deeply embroiled in their subject, the textbook that results gives a very clear reiteration of the Movement at work in the classical style. There is a degree of selection even here. Those who dislike the notion of factual texts presented in a manner that is seen—by them alone—as being terse, aloof or even illiterate, are unlikely to wish to continue further with the course. So we may eliminate yet another undesirable element from the student body.

Second, there are lectures. Basically the information given here is exactly the same as that found in the textbook. Students listen to the words, take copious notes and learn them religiously for the examination; and they find if they are conscientious that they are a restatement of what they have already read in their textbooks.

Some rebel elements decry even this duplication, forgetting that:

(1) this is a sensible and quite correct form of Duplicative Training, ensuring that the data have been communicated *in detail*—which is what the examiner will want to see;

(2) it is only right that students, by each buying a copy of the text, should in this way devote a little patronage to the Movement—and the cost of the books to them would only otherwise be wasted on trifles, hot chocolate and pizza; and

(3) possessing books is itself unnecessary; who ever heard of an Expert without them, may one ask?

In traditionalised 'education' the lectures were of a discursive nature, the aim being to elaborate the student's 'understanding'* of the subject. The study of books out of lecture hours was then seen as a kind of supplementary activity— indeed, students who disliked the term 'education' even then would say that they were *reading for a degree*.

No longer do we hear this as we did. The term 'training' is widely used, and students say they are 'studying' at university these days. As the Expert would agree, *that's more like it*. Meanwhile the enforcement of lecture attendances is rigidly observed; all students are expected to attend a certain number of lectures or they are expelled from the course. To the outsider that may seem heartless: but to the student it must be clear that those who cannot adhere to such a routine are most unlikely to succeed as an Expert.

So quitting the course could well be the *kindest* thing for that individual in the long run.

There are concessions for those who stay the course, mind you. One of these is the distribution of 'niners'—lectures starting at 9:00 a.m. There are plenty of these as a rule in the first year of a course, but as the student gains seniority the number tends to fall. There is thus an interim stimulus to stay the course.

During the lecture itself there are several subliminal aspects of the training process at work, apart from those already enumerated. Students are learning throughout how to memorise, to conform with ease to the system, to obey and to follow; they are discovering how to look alert when (shall we say) a trifle bored and are, by the lecturer's example, understanding much of the Image Projection necessary for an Expert in the field.

In fact the lecture is the most important single factor influencing the student, and the selection of lecturing staff is naturally at the hub of the system. Here we have a superbly well-coordinated example of Nonscience at work.

A principle underlying the Movement is, perforce, randomisation (page 59). To ensure that this applies to the lecturers it is vital that they are chosen for the right reasons, i.e., that they are *not* selected for any lecturing ability *per se*, but for some quite unrelated property. It may be because they have sympathies with the research interests or social position of the head of department (this is one

---

* The repugnance that true Experts feel over this imprecise, woolly generalisation of a term is discussed elsewhere, page 59.

of the commonest randomising factors employed); it may be because of their total number of publications (page 117); or because they are an alumnus of the university and an accredited member of the Past Students' Association—and therefore understands its idiosyncrasies. But they are not, under any circumstances, selected because of mere prowess as a lecturer.

This process is aided by a universal unvoiced agreement on the training of lecturers. Namely, there is none.

In many spheres of activity from soccer to bamboo culture, from egg cookery to space travel, there are standard reference books available and courses of training too. But not lecturing! Experts who become lecturers have backgrounds which range over the entire gamut of Expertistical Endeavour— from routine laboratory investigations to extra-mural administration, some are newly qualified Experts themselves—which reinforces their eligibility.

This sets the seal on the Randomisational Selectivationism, as it may be termed, that underlies the choice of lecturing staff—and to complete the picture there is a careful elimination of feedback in the system which might prejudice us by a knowledge that the lecturer concerned was 'good', 'bad' or 'indifferent'—arbitrary distinctions that they are.

If a class of students turns out to fail generally in the examinations, it is said to be 'a bad year', whereas if the level of passes is very high, the students are said to represent 'a good bunch'. There is no machinery to assess the success/failure proportion in terms of the lecturer's abilities (or lack of them) and, as one would hope in a free society, there is no attempt made by a head of department to sit in on lectures in order to see how well the job is being done.

Thus the lecture embodies most of the essential principles of Nonscience. From the absence of restrictive training or performance feedback to the careful cultivation of Randomisational Selectivation, the lecturer is a veritable keystone of the system.

Third, we have practicals. These are routine sessions of re-worked experiments carried out in the student laboratory and have been neatly described by the Dainton Report (page 58). Feckless and foolishly heterodox individuals are soon eliminated by the coherently planned repetitiousness of these classes, and quite right too.

Fourth, there are tutorials. In these periods small groups of students come together with a member of staff for discussions and suchlike. Students will find here the opportunity to reinforce their training in Pupillary Principles, namely: CONFORMISM, UNINTELLECTUALISM, RAPT ATTENTION MODE, and IMPLICATIONAL SYMPATHETICITY (page 56). Tutorials have, then, much in common with lessons at school, save that the member of staff now has a much better chance to keep an eye on the individual student. It is of vital importance to behave with precision and cultivated correctness according to the principles enumerated. Many a student has been passed or failed

Revolutionary teaching methods recently publicised by the US Information Service. The legend on the blackboard, 'faith in physics', is only justified when the length of the metal nail (L) is less than that of the wooden block by a factor derived from $L = b - (b/10)$.

at examination time, not because of excellence or inferiority in their written answers, but because of the member of staff's opinion. And that is gleaned from the tutorials.

There are, then, many different ways in which the aims of Nonscience are perpetrated and through which they may be brought vividly home to the would-be Expert. Success in all of them is best for a soundly based career in Nonscience. However, the enterprising student can make up no end of leeway by a good performance in tutorials. They are his strength in reserve, one might say; he can correct in the tutorial a whole range of undesirable interpretations of his performance elsewhere by subtle suggestions to the tutor—allusions to sickness in the family, a girlfriend in difficulties, over-conscientiousness or insomnia, etc.

The good student, the potential Top Expert, uses all these facilities to the best advantage.

EXAMINATIONS AND SELECTION

Sitting an examination is really too simple for words. All it requires is a cool head and an iron will to succeed. The student who has:

- attended lectures, practicals, etc.;
- taken notes ostentatiously;
- ensured a promising self/staff tutorial rapport; and
- learnt the notes by heart

cannot fail. On them, and their manifest conformity to the teaching code, rests the future of the entire Nonscience Movement, and clearly anyone who has completed the course deserves to pass.

The more cunning individual, as we have observed, can subvert some of the aims of complete conformism by the added cultivation of extra self/staff contact at tutorials. Though they are not—by definition—as good an Expert-to-be as their more conscientious contemporary, they certainly have a place in the system. Their ability to utilise the best aspects of any traditional procedure to their own advantage is an aspect of Expertism with certain undeniable uses of its own.

It is essential to couch written answers in an examination in the correct terms. Simple instructions in the rubric are easy to miss, but the would-be Expert must *not* miss them. They should not be unduly concerned, for they are easy enough to observe if they simply read them carefully and follow the instructions *to the letter*. Examiners keep the instructions clear and concise, and they follow a set pattern from one year to the next. So you are unlikely to find any sudden surprises.

Basic questions call for a 'description' or an 'account' of some phenomenon or other. They are very easy to answer: one merely categorises what has been consigned to memory. Thus we have a close approximation to the reproduction of information originally imparted through books or class study. Subtler variations utilise instructions such as those found in the rubric, e.g., 'Answer question 1, two questions from questions 2–5 and any one question from questions 6–8' and similar formulae. These are not usually *too* complicated, and (so as to be strictly fair) are generally repeated from year to year in exactly the same form. So there is nothing to worry about there.

But in some questions there is an instruction to integrate the information in some precise manner. The wording may not call merely for a plain description, but terms may be used such as: 'Compare …' or even 'Compare and contrast …'. Set formulae in answering must be adhered to in these cases.

Let us consider some examples so we may systematise the answers required and the form in which they should appear. A *basic* question calls for the mere restatement of facts and data learnt previously during the Indoctrinationalistic Methodological Mode. A *first variation* question calls for the two related sets of comparison; while the *second variation* (compare and contrast) calls for a statement of the salient facts, followed by alternate listing of the points at variance.

Indeed, the order required is not as complex as it may seem, since factors of contrast can easily be made into factors of comparison by a simple re-wording. Thus *Amoeba* and any simple alga can be considered together when we bear in mind their single-celled nature, their aquatic environment, their mode of reproduction, etc., etc., but can be made to seem poles apart if we select characteristics such as mode of nutrition, cell-wall structure and so forth. You may consider any pair of characteristics as being 'compared' or 'contrasted' by a slight change of wording, therefore.

Indeed, the choice of syntactical construction is an all-important facet of successful examination-sitting. The correct phraseology is vital. Examples of some stock interpretations compatible with that to be expected of the Expert and to be incorporated in examination answers wherever possible are given on the next page.

Finally we must emphasise the need to adhere to the syllabus. This applies particularly to the practical examination. We may illustrate this by reference to an actual examination which took place some years ago.* Comments as to the heinous nature of the main departures from Conformistic Normalcy are made at the end of each narrative paragraph.

The examination was a test of practical chemistry analysis, and the object of the exercise was the identification of the metal present in a sample supplied in the form of a ball-bearing.

This is basic enough. In such an examination the student is allowed to have access to such books and analytical tables as they need; the test (in the

---

* Any implication that the author was in any way involved in this episode would be actionable in the extreme.

Concept	Correct means of expression*
I think some other metal does this, too, but I cannot for the life of me remember what it's called.	*Phenomena comparable with those described have been reported to occur with other elemental metals described in the literature.*
I'm not sure that it happens *always*, though …	*The correlation of these parameters should not be taken as automatically invariable.*
What on *earth* do I do next?	*The procedural techniques of routine investigations are then followed.*
I forget the other plant that does it.	*Related characteristics are found in other species that have been documented by previous workers.*
I am too tired to write any more.	*Further evidence in favour of the contention is legion, though clearly beyond the scope of a representative essay.*
To be frank I do not know whether the statement is right or wrong.	*The consistency of these experimental data is interesting, since it lends support to the thesis advanced. However, it would be erroneous to ascribe any definitive accuracy to predictions of such an essentially pragmatic nature.*
It's all done by something whose name escapes me for the moment.	*Further procedures are carried out utilising apparatus/organs/equipment/cytoplasmic inclusions specifically evolved for the purpose.*

* Slight variations of wording are possible in order to obviate the possibility of repetition.

tradition of Nonscience) is an assessment of recipe-following abilities and mute obedience, and access to the literature removes extraneous pressures from the examinee.

The student was given such a sample, which they set down on the bench. Unfortunately, due to an unnoticed slope of the surface, the ball-bearing rolled sedately downhill and landed with a loud 'clunk' in the china sink.

This immediately attracted the attention of other students and of the examiner. It tended to introduce an air of hilarity into the proceedings which immediately turned the examining official against the examinee. Already the test was almost failed.

The metallic sphere then disappeared down the plughole.

The loss of rapport that resulted from this tragic event can scarcely be imagined.

The student then asked for a replacement sample.

The examiner was becoming irritable and his opinion of the examinee and his ability was by now undergoing a subtle reorientation, not altogether in the examinee's favour.

The learned official supplied a replacement. But with an air of entirely justified condescension, he inverted a bench tripod to make a triangular enclosure on the bench surface, by holding it deftly with thumb and forefinger near the base of one leg, and quickly flicking it over. He then popped the ball-bearing into the small enclosed area that resulted.

For a moment (and only for that, as we shall see) the examiner had by this move regained some control over the situation. His reacquired notional superiority improved the chances of the examinee. Temporarily ...

Sadly, unbeknown to the examiner, this tripod had been standing over a lit Bunsen burner for some considerable time beforehand, and the supporting triangle had become almost red-hot. As a result, now that it was settling into the bench surface, dense and acrid clouds of smoke and fumes began to emanate from the area and bright, crackling flames burst into life around the hot metal.

As the coughing, retching students began to fight their way to the door, away from the spreading cloud of fumes, the examiner (who had a weak chest) was unable to retain control of the situation, and left the laboratory hurriedly for fresh air. In allowing an undesirable outcome to become manifest from an inadvertent action on the part of the examiner, the student had reduced his chances dramatically.

After extinguishing the blaze and dousing the hot metal, the student observed that the spherical metal sample was rusty. Only one metal, he recalled, rusts.

He had already settled on the notion that the sample was composed of iron. Because of some superficial characteristic peculiar to this metal, he felt it probably unnecessary to proceed further with the testing procedure. He forgot that this is what he was being tested on! But read on ...

The student called for a magnet (as the class was then reassembled).

Students do not ask for magnets, since they are not taught to misuse them in this manner. Dire results are now inevitable.

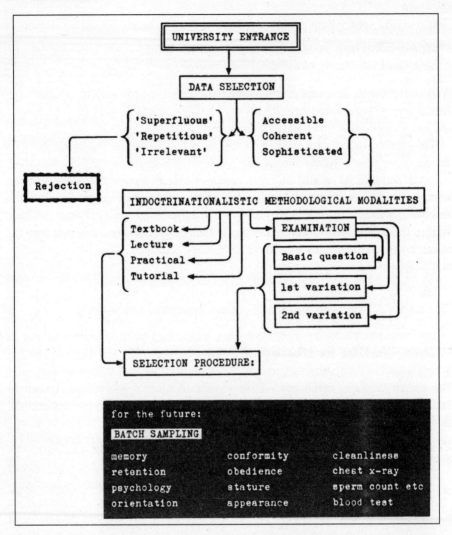

The interrelationships between training and selection are complex. It is important for the would-be Expert to steer clear of outdated traditions that will only result in the rejection syndrome (*q.v.*) and to adhere to the set procedure. For the future there is always the possibility of batch-sampling, such as is used in other industries already. The mass production of Experts is now feasible!

The student observed (and so did the other students) that his ball-bearing was strongly attracted to the magnet, and stuck to it with a loud 'clang'. Indeed, it was difficult to prise it free. This confirmed the diagnosis. The metal was iron.

The student thought he could skate around the standard procedure by this form of jiggery-pokery. What next? He had been taught a perfectly straightforward set of standard manoeuvres which would get him to the answer in his own good time. But no, he had to try to short-cut the whole business. In doing so, of course, he threw away his chances of success.

In that case, the student was destined to failure. The question had been: 'Identify the sample of pure metal supplied using the equipment available.' Of course, one might argue that this had been done. And it had.

But—the conscientious student would have recognised at once that *behind* the question was a subtle implication; a call to Conformism! Had he been properly trained, he would at once have set off on the prescribed pathway towards the self-evident answer. But his mind was too lax by far; he had not benefitted sufficiently from the mind-training that had been offered. And in finding this cheap, irreverent short-cut to the answer he was, clearly, omitting virtually all the attitudes and inclinations for which he had been so painstakingly trained.

He failed, of course; and was no doubt well aware of the reasons. Those who are content merely to find the right answer cannot become Experts, after all.

FUTURE TRENDS IN TRAINING AND SELECTION

The use of teaching machines will no doubt become an important means of training Experts in the future. Not only does the machine ensure that the student has the proper deep-seated orientations, but it is free of the individualism inherent in any personal staff/student relationship. The true unbiased, clinical nature of the system is thus retained.

There is a trend—a satisfying trend, one must add—towards consistency of output as a result of modern training methods which the teaching machine would help still more. This poses an interesting parallel with the assembly line production techniques of modern industry. *Why not batch-sample students?* Coming from regulated backgrounds, carefully monitored during processing and sifted by the selection procedures, the student's final graduation machinery could thus be greatly simplified. Sample students (one in every 20 or 50, say) would be selected and tested exhaustively.

They might go through a range of assessments, including memory, retention, psychological orientation, conformity, obedience tests and so forth. To obtain further information they might be screened for statute, appearance, cleanliness, chest X-ray, sperm count or menstrual irregularities as the case may be, through blood test and exfoliative cytology. The result would be a detailed batch report; in short, a far greater number of parameters could be monitored by this means than have ever been possible before. Rates of throughput could be increased; colleges could be utilised for training on a full-time shiftwork basis with correspondingly greater economies, and automation of the process would complete the picture.

Indeed, we are probably within striking distance of the assembly line production of Experts. The potentialities are limitless.

———————— —m— ————————

LEAVING HOME

Youngsters find that going to University is the easiest way to leave home. Teenagers are meant to leave the parental house (well, if they can afford it, which is not guaranteed these trying times). Parents raise adults, not children. Living on your own is your destiny, but parents are heartbroken if you say so. For example, suppose you found a company in Oxford offering management training in your favourite subject. You'd pass the interview, check out the town, find a cosy bed-sit and locate the nearest pub. It all works perfectly—*until you tell your parents*. They are horror-struck: 'Why are you walking out on us? What've we done wrong? Do you hate us all? Your gran will be heartbroken!' It is a trauma from which the family might never recover.

Now try it another way: be a student. Apply to go to Oxford Brookes University. It isn't anything to do with the world-renowned Oxford University, but your family won't realise that. It's far easier to get into. They ban lecturers whose attitudes they don't like, so you will be shielded from controversial views and, although the accommodation is miles away from the college, a room is allocated for you so there's no need to search for a home. This time the reaction at home is very different: your parents ring gran and gush down the phone: 'Yes! Off to university! Isn't it amazing? We're thrilled!' After three years you are £65,000 in debt and are hunting desperately for a job (remember, there are supermarkets in the town looking for till operators), but at least you have a bike to get around. The family is bursting with pride.

In the previous scenario you have money in the bank, foreign holidays, a crisp company car, and a career path, and three constructive hobbies. Yet still gran won't smile.

DARWIN IS GOD

People raised eyebrows when the original book suggested that Charles Darwin was an Expert. But he was. Everybody worships the man. People say that Darwin was one of the greatest minds in science, the man who showed us how species originated, the man who coined natural selection and who, as naturalist on *The Beagle*, dreamt up the theory of evolution which explains how every species originated. That's what we are all taught. None of it is true. The idea of evolution was 2,000 years old before Darwin was born. It had been taught by Empedocles, Aristotle and Lucretius. Charles Darwin is famous for 'survival of the fittest'— although the term was not in the *Origin of Species* and wasn't even Darwin's phrase (it was coined by Herbert Spencer).

Here is Charles Darwin's idea spelt out: 'The strongest and most active animal should propagate the species, which should thence become improved.' That's the theory, in a nutshell, yet those words were published before Charles was born. They were the conclusions of Erasmus, his grandfather, published in 1794. That same year Richard Sullivan wrote: 'The natural tendency of original propagation to vary to protect the species, produces others better organized. These again produced others more perfect than themselves.' Back in 1745 Pierre Maupertuis had written: 'Chance ... produced an innumerable multitude of individuals; a small number found themselves constructed in such a manner that the parts of the animal were able to satisfy its needs; in another infinitely greater number, there was neither fitness nor order: all of these latter have perished.' There were at least 10 people who published the idea before Charles Darwin. The first evolutionary tree was published by Edward Hitchcock, 55 years before the *Origin of Species* appeared; and indeed Darwin didn't start writing his book until Alfred Russel Wallace had written up the theory and sent it to him by post. Some 27 years earlier, the theory had been presciently published by someone Darwin didn't know— Patrick Matthew: 'Those individuals who possess not the requisite strength, swiftness, hardihood, or cunning, fall prematurely without reproducing ... their place being occupied by the more perfect of their own kind.' Darwin omitted mention of these earlier investigators when he wrote his book. Matthew, on reading Darwin's words, was horrified and he complained. Charles Darwin wrote back: 'I freely acknowledge that Mr. Matthew has anticipated by many years the explanation which I have offered on the origin of species under the name of natural selection. If another edition of my book is called for, I will insert a notice to the foregoing effect.' He didn't. Three editions of Darwin's book came out before Matthew's name crept in—and that is the secret of Darwin's success. His theory wasn't original, but he didn't say so. The earlier publications had caused growing interest in evolution, so that—by the time *Origin of Species* appeared—everybody wanted to know more. That's the rule: Fashionism is what matters. Not originality. And certainly not integrity.

THE UNIVERSITY INDUSTRY

Precisely as predicted, universities have gone on to become big business. The proportion of pupils entering A-level at 16, to prepare them for university two years later, has rocketed up since *Nonscience* was published in 1971. Not that they're

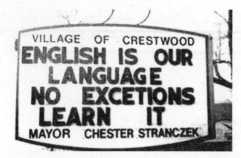

VILLAGE OF CRESTWOOD
ENGLISH IS OUR
LANGUAGE
NO EXCETIONS
LEARN IT
MAYOR CHESTER STRANCZEK

Universities that 'devalue' teaching

From Dr Pete Dorey

Sir, You report (May 29) that the system of ranking and funding university department research is to be reformed. It is a pity that the whole system is not going to be scrapped.

The pressure on academics to conduct research has become so intense, due to the prestige, rankings and funding which research attracts, that university teaching has become devalued and disregarded. Many, if not most, university academics now try to do as little teaching as possible, or even attempt to avoid teaching entirely, so that they can maximise time for research.

University lecturers who do still take teaching seriously are sneered at by their colleagues, or are liable to be told that they are "not a serious academic" (as I was last year, by a professorial colleague).

Now they're weeding out middle-class doctors from medical schools

By Bernard Ginns and Matt Nixson

A FURIOUS row erupted last night after it emerged that universities are being 'bribed' by the Government to accept trainee doctors with sub-standard A-levels.

REJECTED: Zoe didn't get an interview

'Insult' for star pupil who dreamed of being a doctor

HIGH-FLYING Zoe Ganjavi had dreamed of being a doctor since childhood and was expected by her family and teachers to sail into medical school.

As head girl of Stratford-upon-Avon Grammar School in Warwickshire, her academic record was enviable.

The 18-year-old vicar's daughter had already achieved four A grades at AS-level – with 100 per cent in three maths papers – and was predicted to get four As at A-level.

But when she applied to Nottingham University medical school last year, she was not even offered an interview. It was a terrible blow for her and last night her family said Zoe had been discriminated against because she was middle-class and had gone to a good school.

Her father, the Rev John Ganjavi, said: 'I was very angry – I was brought up to believe you got rewarded for your hard work. It's a terrible insult.'

Zoe said: 'Someone who has less than three Cs at A-level will struggle to do well in medicine.'

She gained four A-grade A-levels as predicted and won a place at Southampton University medical school.

Experts know that simple-minded members of the public must be kept in their place. Patronising them (and lowering standards) is the best way to keep them happy.

Daily Mail, Thursday, September 25, 2003

How to eat a carrot

Munch from the bottom and discard the top, the health police tell primary school pupils

By Laura Clark
Education Reporter

TEACHERS these days are used to receiving a barrage of official bumf telling them how to do their jobs.

Never before, however, have they been given a step-by-step guide on how to eat carrots.

Yet that is what has just happened in primary schools across Greater Manchester in an 'urgent' e-mail alerting them to deliveries of fresh carrots for pupils.

A misspelt message from a health official at Salford Primary Care Trust says the best way to eat a carrot is to 'wash it, eat it from the bottom then discard the top'.

The bizarre instructions accompanied boxes of carrots sent to hundreds of schools as part of a nationwide £42million lottery-funded scheme to encourage children to eat fruit and vegetables.

Teachers yesterday said the guidelines were patronising and bureaucracy gone mad.

The e-mail from Ruth Ward, area co-ordinator for the scheme, read 'Please would you forward to all headteachers of primary and infants schools. Urgent.

'Carrots have been delivered this week, the recommendation from the Department of Health is to wash them, eat them from

'We threw it in the bin'

the bottom, discard the top. Any comments use the 'Feedback' form or the school headed paper and post to me.

From Ruth Ward National School Fruit Scheme

Carrots have been delivered this week, the recommendation from The Department of Health is to wash them, cut them from the bottom, discard the top.

Any comments please use the "Feedback" form or school headed paper and post to me, PO Box 149, Winsford, Middlewich CW10 9WW

Food for thought: Part of the e-mail sent out to primary schools

were supposed to stop eating, they just told us to start at the bottom. It's bureaucracy gone mad.'

Sue Moscow, head of Orrishmere Primary in Cheadle Hulme, called the e-mail ridiculous. She said: 'We threw that straight in the bin, it was unbelievable.'

Graham Brock, deputy head at Springwood Primary in Romiley, added: 'They wouldn't send out instructions on how to eat an apple, so why how to eat carrot?'

Last night, Sylvia Chester, regional co-ordinator for the scheme, said: 'I hope our instructions on what to do with the carrots didn't come across as patronising to the teachers, but

sometimes we have to give people a full explanation.'

She added that distribution of the carrots has now been suspended after some complaints about their quality.

The Department of Health said the primary care trust had sent out the e-mail in response to requests from teachers.

'Some children don't have any contact with fruit and vegetables outside school,' said a spokesman.

'As carrots were a new addition to the scheme, some teachers asked for advice on how to make them appealing to pupils. This advice was in response to that.'

l.clark@dailymail.co.uk

15% is pass at maths GCSE

Geraldine Hackett
Education Correspondent

CANDIDATES in this year's GCSE maths exam were able to achieve a grade C pass after scoring only 15%, it emerged this weekend.

Examiners at the Oxford and Cambridge and RSA (OCR) board had to lower the pass mark for grade C because 16-year-olds found much of the examination too difficult.

The disclosure will intensify complaints that standards have been eroded. Even with the lowering of the pass mark, only half of year 11 pupils managed a grade C or higher in this summer's GCSE.

TEACHER OF THE YEAR -LATTEST NOMINATIONS-

week's truck bomb exploded

There are too many people who simply lack the basic abilities to step into the world of work
Digby Jones, director-general of the CBI, registers concern over the falling GCSE pass rate

'pupils' any longer. Our modern era likes to dignify a lack of progress by changing the words we use; just as bin-men became Recycling Management Technicians, school pupils are now called 'students' (I have even heard the term used to describe 7-year-olds). Back in 1970 the proportion of 16-year-olds entering A-levels, studying for university, was 17%; now it is about 60%. Then, the liberal arts subjects were popular. The sciences languished around 5%—but now they are booming. Subjects like drama, music and art have fallen 17% over the last five years, and English is down 25%. This all shows that young people are less likely to understand what they're told. Meanwhile, the STEM subjects (science, technology, engineering and mathematics) have attracted 15,000 more students in 2019 than in 2014, a further increase of 6%. The proportion of their teachers who leave the profession every year was 4.58% 20 years ago. It has since risen to 6%. The syllabus is so constrained by restrictions, precisely as we need in a properly controlled society, that many teachers cannot stand it anymore. Well, they know what to do if they cannot stand the heat in this kitchen.

Of course, the original book's discussion of examinations, tutorials and practicals is now out-of-date. Exams are now made simpler to ensure that virtually everybody passes. They have to—with league tables in vogue, nobody is interested in what people understand, only in how many of them pass. So (if exams prove to be too demanding) the number of students gaining a pass goes down—and so does your institute's rating in the charts. We can't have that, so the rigor of teaching is relaxed and the pass mark lowered until the success rate goes up again. And look how things have changed! The mention of leeches was a joke then, but, just as predicted, things have changed and leeches are a trendy treatment for amputees, for varicose veins and even in beauty salons. What goes around ...

My idea from 1971 that 'teaching machines' would take over has come true in spectacular fashion. There was uproar in May 2020 when British universities announced that they were cancelling all lectures for the next year because of the virus. In many universities you'd hardly notice the difference because lectures are rarer than they were. Nobody minds if you don't go, so many students just stay in bed. Digital access to online teaching means that most lecturers rarely meet their students and wouldn't recognise their names if they did. Teaching yourself is encouraged for today's students. Since you can do this online anyway, without enrolling at university, people wonder why they pay so much for so little. But remember: the whole point of going to university is to get away from home, to do what you want, and to be indoctrinated into patterns of behaviour and conformity that guarantee success and power in adult life. This is what it costs. So, pay up and go (or, if the course has been curtailed because of some pandemic, don't go, but pay up anyway).

CHAPTER 5

A Career as an Expert

Once graduated, you are set for your great adventure into the unknown wide world. And at once there is a choice to be made: do you leave the cloistered, secure environment of the educational establishment, or do you remain snug and safe in the fold?

Let us be honest. Large numbers of people engaged in the training process, as I have repeatedly observed during my studies of the development of Nonscience, are simply not ready to take off on their own by the time graduation day has arrived. A graduate student of (say) 22 years of age has spent 17 or more years in the institutions of the Establishment. But even this may not be enough entirely to overcome the effects of the first four or five free, untrained, uninhibited years. Indeed, some individuals are simply not fitted, by innate ability, to take off on their own. For them the course is clear.

These people must enter what is called postgraduate research. It involves continuation of the training process within the seclusion of the college premises, or at least under close supervision. Often the student is required to prepare a thesis, and they would be well advised to give this matter careful consideration. The dictionary defines a thesis as a *dissertation*. But it defines a dissertation as a *thesis*, which does not help unduly.*

In traditional terms a thesis was an extended elaboration on a theme—a discussion from predetermined premises. However, this is clearly a little too old-fashioned for the clear-cut precision of today's Nonscience. The abstract criterion of 'intellect' is, as we have seen, out of place alongside such hard-and-fast, immutable parameters as length, use of correct terminological constructions, bibliographical content, the external examiner, their beliefs and aspirations, and the way they feel at the time that they come to mark your effort. A bad lunch or a miserable evening at home can sour them enough to

---

* This is, however, a perfect example of Nonscience subtly infiltrating itself into the broader pathways of literature.

tip the balance against you. So it becomes important to cultivate the right relationship with them in the first place. Find out who they are, where they live,
and what their interests are. Do not, for example, refer to blood-curdling animal
experimentation if they are a spare-time anti-vivisectionist; steer clear of
progressive implications if they are well known for conservatism. Above all,
make quite sure that you know and understand their work so that it may be
referred to—obliquely and complimentarily—in your own.

Second, aim at length. Set your sights on the upper limit, i.e., a thesis in
the 8–10,000 word range should be around the 10,000 mark. (Incidentally, it
is vitally important to steer clear of the time-honoured humorous gambit of
writing a thesis eight words long, and explaining that technically it comes
within the range.)

Third, keep clear of argumentation. By this time in your career you are
unlikely to indulge anyway, but controversy and foolish originality can lend an
altogether precocious air to any thesis.

The choice of subject is important here, too, of course. But in this case you
will find that it is often predetermined by the head of department and since it is
your aim to remain there for some time (and to obtain kindly references when
you move on) it is now wisely traditional to accept their suggestions with grat-
itude and no little humility.

An important finishing touch to any thesis is the bibliography. It is an
error to be too smart about this. It is hardly ever taught in colleges that there
are correct abbreviations for bibliographical references, and occasionally students find out the secret for themselves. *But it is a mistake to use these methods.*
Too easily can it look like being clever. Settle instead for the more acceptable
colloquial form of writing the references. Thus instead of writing something
like:

Snodgrass F. J. (1971): Taxation trends—a Prognostication, *J. Internat.
Econ.*, 21 (iv) 144–153;

settle for something altogether more chatty, such as:

Article on 'taxation trends' by Professor Snodgrass, in the *Journal of
International Economics*, Summer of 1971 (facing Esso advertisement).

This shows keenness, without betraying any signs of overt ambitiousness,
which at this stage could be harmful.

And finally try to be comprehensive by bringing in all the main points that
come to mind. The examining Expert will be quick to mark you down if they
find obvious elementary omissions in the work. The safest way to get round this
is to make sure that the thesis is carefully read by someone in the know on such
matters. If they make suggestions that *this* aspect is neglected or *that* should
be included as well, the remedy is simple. Insert a throw-away reference which

shows you thought of it first, but write it in a manner that does not encourage a deeper inspection, thus:

'*So-and-so*, which might be considered to exert some tangible weighting to the results, has been ignored in the interpretation of the present experimental work'

or:

'*Such-and-such* factors have been eliminated in order to reduce the inconsistencies incurred by the incorporation of too many unknown variables.'

That will generally do the trick.

Above all, do not worry unduly about the success of the thesis. If you play your cards right, attempt to meet the examiner socially or write (as though not knowing that they are going to mark your paper) and ask for advice in a polite and subservient manner, and maintain a correct relationship with the head of department, then a 'pass' is virtually a certainty. After all, the success rate at this level is taken as an indication of the efficiency of the department, and no head would want to create a bad impression on that score.

Postgraduate students must be determined in their resolve to succeed. It has been shown that about one-third of British postgraduate students, looking back on their course, would have preferred them to have been different. Take care not to join the ranks of these lower grade individuals.

Teacher training is the avenue followed by many aspiring Experts who choose to stay in the fold (as one would expect when the potentialities for training senior students are so exciting); a career in school teaching is regarded as generally inferior to other forms of activity. The true student of Nonscience is willing to fall back on teaching if all else fails, and this itself tends to aid the aims of Nonscience in schools. It is the conditioned obedience of the Expert which we need in our colleges if the aims of Nonscience are to be promulgated, when all is said and done, and the less adept individual can do well in working lower down the training ladder. It must be emphasised that the inconsistencies so introduced into school training methods are often good grounding for the scholar in what lies ahead.

Suitably enough, teacher training does not involve a consideration of matters of mere worldly consequence; and one may be confident that such subjects (how to use the tube, how to complete an application form, what is a mortgage, etc.) are left obscure in the mind of the student. But studies of masturbation in its many forms, coupled with practical experience at tea-making, ensure a thorough grounding in the attitudes of Nonscience—ready for them to be passed on to the next generation.

Many of us have heard occasional newly qualified teachers (and, perhaps even more so, senile members of senior teaching staff) suggest that modern methods of teacher training 'have no relevance whatever to the modern school and its occupants.' If only these short-sighted individuals could look below the surface! If only they understood that Nonscience is almost founded on this very fact—it is not 'irrelevance' at all, but *careful randomisation.*

And finally there is the alternative of leaving the training environment altogether and, as it is called, *getting a job.*

This invariably poses problems for the newly qualified Expert. They have all the conditioned, peak performance abilities that their paper qualifications show—yet many employers do not (or will not) recognise the fact. In the industrial sector particularly it is felt that the incoming Expert suffers from drawbacks—lack of contact with the working environment, inexperience, failure to understand problems, etc.—which it is clearly up to the employer to rectify if they see fit.

But (and this is a question that disciples of Nonscience must ask themselves) *is it necessary to change at all?*

The true benefits of having Experts in the firm are more basic. It is they who can make profound statements on matters of controversy. It is to them that the directors should go when a weighty and abstruse document has to be written to confuse a client. Experts are the source of pronouncements—and they may feel strongly that the wishes of some employers (who want to see practical skills and experience) would reduce them to the rank of a working person who ought to know these things.

It is partly because of this state of affairs that the so-called 'brain drain' has arisen. Throughout the 1960s, British Experts tended to leave such a restrictive society and seek recognition and appreciation elsewhere.

It would be wrong for us to assume that Britain has a particularly notable record in this respect. It has been shown previously that if we draw up a list of European nations showing a 'league table' of countries losing a proportion of graduates to the United States, Britain comes about two-thirds of the way down. And anyway, over half the respondents to a questionnaire in 1970 who have been abroad to the United States and Canada had 'already returned', so emigration is not necessarily permanent. Statistics generally available suggest that less than half of all postgraduate qualified individuals who go abroad stay there.

So what lies in store for the vast bulk who remain behind in the land which fostered them? There is evidence of dissatisfaction in industry, not only from the employers (who surely have only themselves to blame if they have not yet adapted to the Expert's basic needs) but from employees too. Evidence suggests that around one-half of all such people employed in industry are dissatisfied with their posts and would either like, or are actively seeking, alternative employment. So we must obviously pay special attention to choice of a job.

Posts in governmental offices are relatively secure. Unless some act of great severity is committed (rape, theft, murder, etc.) it is virtually impossible to be given the sack. Dismissal can come about through the abuse of power and the voicing of controversial opinions, mind you, so it is best to err on the side of obscurity and caution rather than risk being put onto a pinnacle. But it is quite an easy matter, once in, to enjoy a secure and comfortable existence, with a reasonable salary and good prospects of perks (travel, etc.), with very little effort.

Large and expensive gifts of apparatus are sometimes made by companies with an interest close to that of an Expert. One example of the kind of thing you can acquire is shown in the figure below.

The avenue to explore is one where results are not really important: i.e., where the problem is well known and self-evident, and therefore where a break-through is not considered likely. Pest control is one example, cancer research is another. For those who prefer to work a little more in the public eye, there are many fields which are now considered 'newsworthy' by the mass media (though they are no more important than vast numbers of similar fields of work which are never referred to). Small items of progress in these fields are inevitably widely reported in the press and on radio and TV, and the Expert is assured of a regular casual income from such sources. The Pill and pollution head the list in 1970 with cervical cancer and prostaglandins a very close second. Then

A novel array of apparatus which appeared in the correspondence columns of *New Scientist*. It was built to investigate 'the nature of randomness', from which one's own conclusions may be drawn.

Space research, now sadly losing Fashionism, has been a regular employer of Experts on a large scale. Their enviable working conditions are shown in our photographic study.

there are fallout (which has lost a great deal of public appeal, and therefore assures a more restful career) and spina bifida (which must be mispronounced properly—as *biffider*, not *by-feeder*—if it is to be socially acceptable) along with astronomy (very useful, and very newsworthy) and a host of similar fields of research. Pollution is at the moment reaching a temporary peak of public attention, and there are certainly gains to be made here. The prospects of being widely quoted are good, and authorship rights should be promising, even if the subject loses popularity again soon.

This example does point to the care that must be taken in selecting a career. Some technologically orientated (or *oriented*, as the Americans will say) fields demand a considerable amount of hard work which the salary may well, quite simply, not warrant.

Lasers are an example here, and so are computers. But many others are destined for a fairly predictable rise and fall from grace. High in 1970 are the Pill and pollution, certainly; but viruses should be watched with care and the whole field of metaphysics (e.g., ESP, telepathy, mood control and memory transfer) looks as though it might have something to offer before too long. Nuclear research and antibiotics are both *passé* at present, and cancer is probably past its peak.

Epidemiology of human diseases is a Fashionistic field at present. Spina bifida is a splendid case in point, for children suffering from this condition are produced at a fairly predictable rate and can, with surgery, be encouraged to survive for as long as possible, sometimes even to the point where they breed successfully. This is important, of course, if future research material is to be provided (since the condition is primarily genetic). Meanwhile very large amounts of money are forthcoming from government and public sources for research into just this kind of condition, which makes the job secure. At the same time, teams of social workers are employed to study the effect of the deformed child on the family that have produced it.

Thus we have a kind of self-perpetuating system in which the role of the Expert is fundamental. But in a case like this, where progress and practical results are both unlikely, and where output of work is therefore at a minimum, competition for jobs is high. But this is an example of what *can* be attained.

However, there is one cautionary note which has to be added. It's all very well aiming at such a post (and eventually planning to move towards the occupations for Senior Experts discussed in Chapter 11) but there is a chance that, even before you reach this stage, something can go wrong. You may never obtain a post in the first place.

This is often because of the use of some predetermined Blanket Criterion used by the selection committee to whittle down a list of applicants to sizeable proportions. Quite often, nationality or race may be used (in this way Experts with foreign-sounding names were never, but never, given an interview in the 1960s—one I came across claimed to have submitted 'thousands' of forms without ever having any luck). Religion, though now not Fashionistic in job selection, is still found occasionally. But the commonest arbitrary hurdle is the Level of the Degree.

What happens is this: all applicants with anything less than a first-class honours degree are automatically cut out of the list; this leaves a group which is far more manageable in size. The difficulty is that most of the 'plum' jobs are widely advertised, and therefore are snowed under by applicants. In this way anyone with a second- or third-class degree is never even considered for a post, no matter how suitable they may be. These are the people who go around for years on end trying to find a job, without utilising the criteria outlined on page 106. They are stuck with their degree, of course, and cannot seem to find a way out of their predicament.

In these cases the most efficacious answer is the technique known as Notional Upgrading. This means that the applicant states on their application form that they have a first-class degree, whatever the class they may actually have.

At once they become eligible for consideration, and may well land the job. No-one ever checks back to source in matters of this kind, and so the discrepancy is unlikely to come to light. If it does, it is a simple matter to point to the obvious slip of the hand that had led to the error in the first place (i.e., by suggesting that you put down the wrong grade in your haste to apply, or perhaps that the college authorities gave you a second—instead of a first—as a result of their own failure to recognise talent when they saw it). 'Is this not', I can hear the perceptive reader ask, 'lying?' Not quite. If a typist puts down 120 words per minute on their application form when they can type only 80, they will at once be found out and seen to be a Perpetrator of Falsehoods. State that you can speak French when you cannot, and you will be quickly uncovered as a charlatan. Claim to have a silver medal at ballroom dancing and you can be immediately challenged if in fact you are a grade or two lower. In bricklaying, plastering, playing a musical instrument, running, design, bank management and so many other more menial fields an inflated qualification would be rapidly deduced.

But the essential randomised nature of the Expert's paper qualifications means that this cannot be done in Nonscience. Another way of looking at it is to say that mere Lay Persons haven't the perception to notice it.

Experts, remember, are simply out of their Class.

At the basis of all sponsored research is the Grant. Grants are great fun. They consist of very large cash sums given to a research worker. The amounts are decided by a committee. Happily, this body is rarely in close contact with what is actually going on and therefore the amount of money in the grant bears a closer relationship to the determination and persistence of the grantee rather than to any sanguinity in the granter.

Almost all grants are determined *not* by the amount actually required, *but by the sum granted last time*. Now this is important. It means that any sum granted simply has to be used up in the allotted time, or else a smaller amount will be forthcoming in the next round. It is therefore important to start off as you mean to go on, with a very large sum of money to play with. As a general rule it is the international organisations—the WHO, the FAO, etc.—where truly excessive grants are given and where grantees often say it is difficult to find ways of getting rid of it all. Government organisations (the Medical Research Council, etc.) are quite beneficent in this respect too, while the scruffier colleges generally have to coax money from local industries.

The use of a grant is a special study in itself. It is naïvety of the most elementary kind to assume that money saved from one year can be used in the next, so let's dispose of that right away. Perhaps I should recall some personal experiences in this regard which will show how patently foolish it is to go into these things unprepared.

It occurred in a government laboratory where, in the most junior capacity conceivable, I was going to spend a year before continuing my studies at university. One of the first tasks put before me was to draw up a list of materials and equipment that I would need during the following 12 months. And this I did. A litre of this, two dozen tubes of that, a yard-and-a-half of the other ... the reader will, I am sure, be able to imagine the kind of thing.

Eventually I presented the list to my superior. He grasped the fragment of paper in his knotted, firm hand and his eyes flicked Expertly down the columns. I watched with some interest as beads of perspiration began to break in neat, regimented rows over his forehead, collecting in the laughter lines that littered the landscape of his face and slowly running down in ordered streams like flood water in furrows. He was turning purple. Suddenly he seemed unable to contain himself any longer, and burst out with a bellowing laugh that lives with me to this very day.

At once I realised that I had overstepped the mark with, it appeared, somewhat amusing results. I went with him into his own room while he totted up the figures, quivering with a chuckling laugh from time to time. The total came, I observed with a growing feeling of horror, to over £400—and all for one small laboratory!

I began to go over in my mind where cuts might be made—perhaps we could do with only a yard instead, and so forth—when all at once he spoke. And then I realised that I had not fully grasped the realities of the situation.

'Good heavens, lad,' he expostulated, quivering, 'four hundred quid? For your laboratory? The standard grant is *two and a half thousand!*'

I ventured to explain, searching for justification but only getting myself in deeper as I went, that we could save the extra £2,100 for next year which (as the perceptive reader will by now be anticipating) reduced him to near-hysteria.

'If that's all we asked for,' he gasped eventually, 'that's the most we would ever get again. Not only that, but they'd be wondering what the devil we had been doing with all the extra in past years, if we hadn't actually *needed* it.' There was by now an urgency in his voice, which, at the time, I could not quite understand.

Now, I trust, all is clear. The principles of Nonscience put the matter soundly into perspective. Thus, when applying for a grant, bear in mind the following points:

(a) It cannot under any circumstances be smaller than last year's. Not only will you be in receipt of a dramatically smaller grant next time, but *questions will be asked.*

(b) Do not try to save money for the following session. If you need, say, a calculator unit then buy half now and the other half later. Money meant for 1972 is meant for just that, and cannot possibly be saved for 1973.

(c) Invariably give estimates of overspending, rather than risk erring on the other side. In this way a claim may be trimmed down to match the grant total and the committee will feel that they are, if anything, doing you out of your rightful needs. You are left, thereby, in good stead for a future claim.

(d) Always ensure that consumables are well highlighted. This is for two reasons:

    (1) An application for more capital equipment has a ring of laziness about it. The committee will be wondering if the team has been installing too many labour-saving devices and is meanwhile being left with time on their hands.

    (2) Consumables sound as though a lot of work is under way. The committee may even be inclined to give you some bigger and better apparatus to use them up in.

    Always, therefore, aim to have a rapid turnover of materials such as platinum wire, alcohol, benzene, etc.; technicians may be able to help with suggestions as to how the grant may be reclaimed at its face value. What is not purchased from the department by others can always be used for personal purposes, by helping to boost one's drinks a little, or by economising on petrol.

(e) Do not allow *any* tidying or wiping round before a grant committee is due to visit your establishment (should that rare eventuality arise). They may feel that the generally degenerate air marks a laboratory where the injection of a little extra cash can help put it on its feet again. And anyway, *a busy lab is a scruffy lab.*

There is no need to stay stuck in the laboratory, mind you. Many Experts have found themselves a niche in a government establishment which provides them with the outlet for power and prestige, without the need actually to meddle with complex and sometimes dangerous apparatus.

One prominent avenue that you might explore is the preparation of official statistics. Vast numbers of opportunities lie waiting for the right people. As you will know, the behaviour of any economical system is carefully monitored by what are called 'official sources'. Thus after a British dock strike a few years back it was stated that 'some £90m had been lost in the export market.' Trading figures, expenditure totals, wage indices and the like are regularly quoted in official publications with great accuracy (quite like the Common Market figures we have seen widely discussed). In fact, they are mainly notional amounts, which means that they are basically arbitrary (i.e., unrelated to figures pertaining in actual fact). But they are accepted without question by the world at large.

Part of this excitingly randomised system lies in the government Expert's power to produce obscurantific literature. Consider this example from the Weights and Measures section of the British official Expert establishment:

'Every letter in any such words apart from the initial letter of such words shall be of such size that the smallest rectangle capable of enclosing each

```
TO ALL MEMBERS OF THE DEPARTMENT
VISIT BY THE UNIVERSITY GRANTS COMMITTEE
TECHNOLOGY SUB-COMMITTEE
THURSDAY, 11th FEBRUARY, 1971

    The above Committee will be visiting the Department
on Thursday, 11th February between 11.15 a.m. and 12.30 p.m.
and between 2.00 p.m. and 3.00 p.m.

    We clearly want to give a good impression and for
this reason all members of the Department are asked to
implement the following:

1. All laboratories are to be cleared up and rubbish or
equipment not used is to be stored away without, however,
giving the impression that any special steps have been
taken to tidy up the place.

2. All research workers are asked to work in the
laboratories during the period of these visits simulating
'feverish activity'.

    The intention is that we should give the impression
of being a highly active Department which looks after its
resources and utilises them to the utmost.

                                         S.A. TOBIAS

8th February 1971
Department of Mechanical Engineering
```

How to alert the laboratory that a grant committee is on its way. The University of Birmingham's student newspaper, *Redbrick*, in reporting the above memorandum, published a suggestion that it betrayed a very 'peculiar' attitude. Rubbish! It merely demonstrated how Professor Tobias was anticipating this book, that's all.

letter of every such word shall not be less than 9/16th of the area of the smallest rectangle capable of enclosing the largest letter, apart from initial letters, in any word of more than one letter appearing on any label on that container.'

Or, by comparison, look at these extracts from a US research report on Educational Theory Models:

'Toputness is system environmentness … and storeputness is a system with inputness that is not fromputness. Disconnectionness is not either complete connectionness or strongness or unilateralness or weakness, and some components are not connected … in other words, feedinness is the shared information between toputness and inputness where the toputness is prior to the inputness.'

The results of all this on the Lay Person are predictable. Early in 1971, a British government department announced that after examining nearly 2,000 forms filled in as part of a development scheme, there was *not a single one* that was correctly completed. Another job well done, Experts!

This shows how we can be sure that official statistics, collected through such means, are invariably unquestionable. If you are told that, for instance, corrosion causes the loss of £55m to government agencies per annum in the United Kingdom, or £350m to transport, you would be inclined to accept this as an immutable, accurate statement of fact. The true Expert, however, bears in mind the essentially notional nature of such totals. Not only that, *but they know why.*

Industry as a whole manages to perform the same kind of manoeuvre. We hear reports of 650 tons of sulphur dioxide released per day from a new power station in Wales, or 400 tons per day from another near Plymouth and naturally Lay Persons are not sure what to make of it. If they thought in terms of a thousand tons of sulphurous acid being sprayed over their home town they would probably object, heedless of the benefits that increased electricity must mean to Nonscience (and therefore, indirectly, to them) eventually.

Instead, we must emphasise the clean nature of the gas (it is invisible, and not therefore responsible for blacking out the sunlight over our cities) and its purity (the SO_2 emitted is extraordinarily pure). It is the Expert, once again, who makes these pronouncements of safety.

Experts in industry often have to juggle with figures, too. Put simply, the world of capital investment runs entirely on what are called 'predictions' (in socialist countries 'prognostications', but they mean the same). These are similar to guesses—and have been called *guesstimates* by certain irreverent people in the past—with one cardinal difference: guesses are made by Lay People, while predictions are made by Experts.

For example, the Mere Lay Person will always try to guess what outlay or expenditure they'll need for a particular product. And a fat lot of good *that* will do them. The Expert, on the other hand, adopts a far more professional approach. In tendering for a contract, for instance, the officially recommended approach for Experts is this:

(a) You will guess that you have a chance of securing it at a certain price—what chance? Let's write them down thus:

95% chance of securing contract at					£95,000 (with tiny profit)
80%	"	"	"	"	" £125,000 (with better profit)
75%	"	"	"	"	" £145,000 (with bigger profit still)
60%	"	"	"	"	" £160,000 (with large profit)
40%	"	"	"	"	" £180,000 (with huge profit)
20%	"	"	"	"	" £195,000 (with vast profit)

(b) Then multiply the estimated profit by the chances of obtaining it and note down the figures.

(c) Finally, select the highest figure.

Another approach is known as *cost–benefit analysis*, and takes factors into account by putting a value on them. Thus for a new industrial development you might value the noise nuisance at, for instance, £55,000 and the aesthetic spoliation at £92,000. This gives you something to work on in considering rival schemes.

Both these methods were widely advocated in the 1960s. Many people suggested that they gave a spurious authority to figures that are only guesses. But this begs the question. Like results from a computer (which has only been programmed by someone's estimates, after all) these are not guesses at all, *for they were not made by Mere Lay People.*

They were engendered by Experts. The world of economics, commerce and industry depends on this very fact.

Joining industry has several advantages:

First, you are where the source of all the money is. It is less complicated, therefore, to divert some of it to your own ends.

Second, the industrialist considers there to be a certain kudos attached to the employment of a quota of Experts, quite similar to the aura of prestige he associated with a computer in accounts, or electric typewriters in the correspondence pool. You become a status symbol, in other words.

Third, quite apart from pay, fringe benefits (perks) are considerable. Quite extensive travel (which can become worldwide if you play your cards close to your chest) is one bonus, and a liberal expense account is another.

One must realise that many departments in the world of industry are maintained solely for reasons of prestige. It is incumbent upon the Expert to keep the laboratory or office looking efficient and busily occupied for the benefit of visiting notables. Actual results, in terms of output per unit staff, or work as a function of expenditure, are not required.

Not all posts are like this, of course. But the clue is to look for the coded information imparted to you at the initial interview of this kind:

(a) 'We are hoping that we can expand this department, one day.'

(b) 'Considerable freedom is allowed to staff in this department, which is still largely experimental.'

(c) 'The scope of research is left largely to the initiative of individual staff members.'

All of which brings us to the common denominator in applying for any job—the application form. No matter if this is for a job in industry, a post in local or central government, a teaching position or one at a college, the same basic rules apply.

In the first instance, *nobody reads them*. Application forms are meant to be filled in and they will be glanced at, to see this condition has been fulfilled; but that is all. They have a traditional niche in the selection procedure for Experts because they demand detailed information of a peculiarly personal kind—and only the true Expert is likely to wade through them with sufficient loyalty. The typical form has to be filled in in triplicate (at least) and occupies many, many pages. When questions include details of dates, places, parental statistics and so on going back over your entire life history they may take days to complete—but it is all part of the need to 'show willing'.

Some individuals feel uneasy about application forms for the rather childish reason that they do not know where the information goes. It seems that some people—qualified Experts, we must remember—dislike writing out a detailed personal history full of intimate and confidential details, only to find it passed around whither they know not. To these people we can only say 'be calm'. The forms are sent to Experts after all; and who would seriously doubt that an Expert would bandy the information around lightly?

It is important to ensure that a Positively Projected Image appears on the application form, for though the sections on dates, schools, examinations passed, experience, etc., are generally ignored by the selection process, the paragraphs on 'special interests' and 'reasons for applying' may well be read through. So consider carefully your reply. Best of all, work at it while still in college, and bang off a verbatim copy of the carefully edited final draft to each and every prospective employer.

Emphasise such Expert-orientated occupations as chess, reading, visiting museums, collecting; but do not be tempted to put pot-holing, rally-driving or pub-crawling unless you are *absolutely sure* that such things are Fashionistic for the employer too at that time. To find out if there are special interests with which you can show sympathy, why not ring the head of the department's office and find out what *theirs* are?

You may be able to collar a student or staff member there, if you live nearby, and simply ask. Or if distance precludes this it is perfectly simple to ring up and make some discreet enquiries. Do not reveal your identity at this time; simply imply you are someone from industry who may be able to give some money, or supply some research contracts. One approach is to ask about a visit, and then add 'Would any time be inconvenient?'

'I'm afraid evenings are out of the question, as Professor likes to attend to his butterfly collection' or 'Weekends are difficult, since that is when the Director usually goes yachting' will soon give you the background you require. It is then a simple matter to make sure that *your* interests tie in with *theirs*.

Above all, resist the temptation to supply whimsical answers to questions such as:

SEX: *Frequently.*

REASONS FOR APPLYING: *Overdraft.*

DO YOU TAKE ALCOHOL IN ANY FORM? *The liquid only.*

When considering what kind of post to hold, bear in mind that you can always change later when you have some experience behind you (Chapter 10). Above all, make sure you *always have at least one* job. This is not as idiosyncratic as it seems. Often there is the opportunity for the extension of a research fellowship, or something of the sort, which you may feel disinclined to accept because you feel it is likely that something better will turn up meanwhile. But never resign any post, even in that manner, until you have the next one securely in the bag.

The prospective employer who is going to consider your application will be much impressed if you can explain that you must give notice to your college before accepting the new job. But if you have left already you are technically unemployed and even if that means you can start right away, he will prefer someone who is already attached rather than someone who is virtually on the dole. You can lie about that, of course, as many people do.

Once you have been left *without* employment for even a matter of a week or so then any organisation to whom you apply will automatically count you out of the running. The result can either be the relief queue, or a thoroughly demeaning job. In a survey I carried out in 1968 we came across people with excellent academic qualifications who were in the following jobs:

Technician (alongside 16-year-olds)	Car-washer in garage
Ward orderly	Labourer (digging trenches)
Temporary clerical assistant	" (for British Rail)
Stamp-dealer's mate	" (farm work)
Rocking horse manufacturer	Carpet fitter
Bread van sales driver	Barman
Shop assistant	Petrol pump attendant
Junior secretary	… and so forth.

Most people get into this rut of unemployment through a failure to hang onto a job already obtained, or frequently through applying for posts for which there is, simply, too much competition. The best kinds of advertisements to follow up are those that:

(a) are in *Welsh*;
(b) appear in one edition only of the *Guardian* or the *Journal of Municipal Effluent Disposal*; or
(c) call for some special knowledge (e.g., the virology of mine drainage gas or the permeability of plastics to rare vapours) that can be mugged up specially for the interview. This is a comparatively rarely used ploy, but can be used with telling advantage. If you as a candidate for a post are able to talk as man-to-man (Expert-to-Expert, one might say) with the specialist on the selection panel, then you stand a good chance of being In. The only effort required is a reading of some symposium reports—a trifling matter—and a general acquaintance with the literature on the subject.

Two personal friends who followed this method recently tell me they are now perfectly settled. Both were on the dole. The one has a large laboratory (with technical staff) to himself with but negligible work to do from one month's end to the next. The other joined an international organisation and when last heard of was basking in the East African sun with a bundle of petty cash vouchers in his case. I am not exaggerating.

Too many people are caught up by the kind of advertisement which is snazzily worded and appears in everything from *New Scientist* and the *Times*

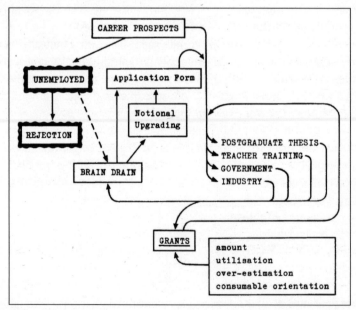

How to avoid the dreadful fate of unemployment and rejection—this diagram should be committed to memory before you proceed.

Business News down to* *Private Eye* and *Oz*. There will be hundreds of other applicants. But choose a post from the above categories and competition is slight, often so slight that the interviewers are glad to have found you at all.

Finally, a word on what happens after you have sent off an application form.

Many advertisements include the words: *'All applications answered'*, or words to that effect. This is, of course, like o.n.o. or a.s.a.p., a well-known abbreviation commonly used for economy of words. It actually is meant to read: *'All applications answered that we decide to follow up.'* The generally accepted way for an employer to inform you, as courteously as they may, that you are not to be short-listed is for them to send you no further communications. It is important that you realise this. So many people, not knowing this elementary fact, have sat and built hopes on an application they've submitted and have actually considered the prospective employer to be *somewhat rude*, of all things, in their failure to send as much as an acknowledgement.

They should realise that in this high-speed technocratic environment there is no time for dithering formalities. It is the crisp, efficient economy of modern conventions which we see at work instead, and the uninitiated had simply better realise the fact.

THE PUBLICATION IMPERATIVE

Experts have to put their work in print. Writing a dissertation (sorry, thesis) has become mechanised and is now much simpler than it was. With the World Wide Web everything is available to everybody, so you don't have to do any work at all. Just look up your subject, cut-and-paste the sections you need, and give them a tweak. That's it. In the television cartoon, Bart Simpson was once complimented by his sister on writing a report. 'The Internet wrote it,' he retorted, 'All I did was hand it in.' Bear in mind that crafty supervisors use software to search for plagiarism, so you need to tweak properly: go through your text and change words in most sentences. A sentence that reads: 'This complex apparatus was assembled and the process began,' becomes 'This *complicated* apparatus was assembled and the process *started*,' so the plagiarism software won't recognise it.

CHOOSING YOUR SUBJECT

You will need to choose a topic, and the more abstruse it is, the better. As the original book predicted, there was a lot of potential in 'slow viruses' and one of

---

* Or 'up to', depending on how you look at these things.

those turned out to be a prion, the cause of Bovine Spongiform Encephalopathy or BSE*, Gerstmann–Sträußler–Scheinker syndrome, Creutzfeldt–Jakob disease, and Kuru. When BSE hit the headlines I was roped into a television discussion about the implications, and in the green room I was discussing the tragedy it would present. One of the leading researchers came bounding over. 'God,' he said. 'This BSE thing is going to be huge. Millions of people could die!' I was about to say how tragic this was, but he chuckled merrily on: 'There is going to be a load of money we can make out of this!' You see? That's how a true Expert thinks.

BSE was (luckily) a one-off, though cases still occur. Most subjects go on and on … cancer research, for instance, is fabulously well-funded, indeed I have spoken to institute directors who privately admit they have more money than they can usefully spend on research. Spina bifida used to provide lots of opportunities until the 1990s, when it was confirmed that vitamin B9 (folic acid) can help prevent it. If pregnant women take this vitamin during pregnancy the incidence of spina bifida is only 30% of what it was. Contrary to what many people believe, folic acid doesn't prevent it entirely, but it does mean there are fewer opportunities to make money. The vitamin should be added to flour, but that still isn't done.

Top of the current charts is the coronavirus pandemic, and limitless sums of money are available. In April 2020 both Imperial College and Oxford University announced they were developing vaccines. This is what Universities do, as a matter of course; it doesn't require much hardware and they already have the staff. Both approached the government for finance … how much, might you guess? £10,000? £100,000? They each grabbed £20,000,000. Superb. Nobody mentioned there were 200 such projects already, with 30 vaccines soon to be tested.

High in the charts at present is dementia. Huge publicity campaigns are urging people to get diagnosed as soon as possible. There are online tests people are encouraged to take. Why? Nobody can cure dementia, and you cannot prevent it; indeed—until you're really ill—there isn't a drug that can do much to help, and even the best have disappointing results. Nobody knows what causes it. Nobody has any idea why it is becoming more widespread. In a nutshell, nobody knows anything much. They love to say that keeping the mind active wards off the disease, but when you think that Margaret Thatcher and Terry Pratchett both succumbed, it is obvious that cannot be the answer.

Once a patient has been diagnosed with dementia, they are put into a neat little box with a label. And that's it. Between us, there is no benefit to the patient of an early diagnosis. Nothing can delay it or ward off the condition, and it compromises their chance of getting insurance—but for Experts it has several advantages. First, it offers a vast supply of people who were recently a bit forgetful (but are now certifiably doomed) which gives you statistics to study. Second, because it cannot

---

* Mad cow disease has just 4 syllables; bovine spongiform encephalopathy has 11. That's the way to go.

be cured, it isn't covered by the National Health Service. This is brilliant! The NHS was set up to cure people of their illnesses free of charge—but, if someone can't be cured, then they have to pay! Obviously, the sooner someone is diagnosed, the better it is for all of us. Except them.

FAKE NEWS AND FORTUNES

Finding a topic for research has become more difficult because of the garbage that appears online. Much of what is stated online is nonsense. Google reveals 540,000 web pages discussing the Amazon rainforest as the 'lungs of the earth', but they aren't. Almost as many sites claim that the Amazon provides 20% of the oxygen we breathe, but it actually produces nothing. These ideas are mistakes people copy until they spread over the web like a fungus. We know of fake news but the Internet brings us endless false facts and today's students can rarely tell the difference. The Graduate School of Education at Stanford University surveyed the reactions of 7,500 students to false accounts on the web. The researchers were dismayed and found they faced a bleak response; the inability of youngsters to tell reality from fiction was found with 'stunning and dismaying consistency'. So where can you find reliable information? Published encyclopaedias? They are said to be repositories of double-checked, objective information approved by experienced editors, and you might fondly imagine that the facts are from authorities in the field, but that isn't the case. Encyclopaedias obtain their information by stealing it from—other encyclopaedias. Once a mistake has been published, it continues down the line. The discovery of Brownian Motion by the botanist Robert Brown in 1827 is wrongly explained in almost every reference book. He discovered this movement in tiny particles teeming inside pollen grains. Under his microscope, he could see that there were thousands of minute specks within each grain of pollen, and these specks were incessantly moving (they're being bombarded by jiggling atoms nearby). By the time it was written up for encyclopaedias, they'd got it wrong: they now said the entire pollen grains were in motion. That's ridiculous—each grain of pollen is tens of thousands of times too big to be jostled by tiny atoms. It makes no sense. Even so, it is stated that way in almost every encyclopaedia. When the BBC wanted to prove the point, they had physicist Jameel Sadik Al-Khalili (popularly known as 'Jim') demonstrate it on television. There he sat, with a little tank of water, and a microscope, and a pot of pollen. He tipped the pollen onto the water and described how it was jiggling about. It wasn't, of course; but he didn't have a clue how to look through the microscope, so nobody realised that the entire experiment was a ghastly failure. And don't blame the Beeb, or even the hapless presenter; if everyone else gets it wrong, why shouldn't they?

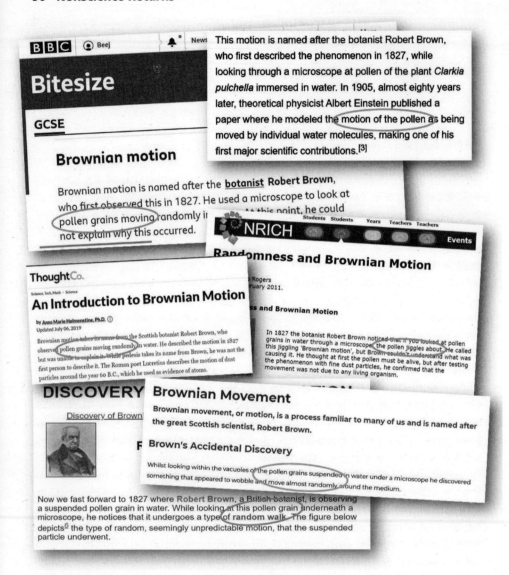

BBC ⊙ Beej 🔔° News

Bitesize

GCSE

Brownian motion

Brownian motion is named after the **botanist** Robert Brown, who first observed this in 1827. He used a microscope to look at pollen grains moving randomly i̶n̶... At this point, he could not explain why this occurred.

This motion is named after the botanist Robert Brown, who first described the phenomenon in 1827, while looking through a microscope at pollen of the plant *Clarkia pulchella* immersed in water. In 1905, almost eighty years later, theoretical physicist Albert Einstein published a paper where he modeled the motion of the pollen as being moved by individual water molecules, making one of his first major scientific contributions.[3]

NRICH ⬤ ⬤ ⬤ Students Students Years Teachers Teachers Events

Randomness and Brownian Motion

s Rogers
ruary 2011.

ss and Brownian Motion

In 1827 the botanist Robert Brown noticed that if you looked at pollen grains in water through a microscope the pollen jiggles about. He called this jiggling 'Brownian motion', but Brown couldn't understand what was causing it. He thought at first the pollen must be alive, but after testing the phenomenon with fine dust particles, he confirmed that the movement was not due to any living organism.

ThoughtCo.
Science, Tech, Math › Science

An Introduction to Brownian Motion

by Anne Marie Helmenstine, Ph.D. ⓘ
Updated July 06, 2019

Brownian motion takes its name from the Scottish botanist Robert Brown, who observed pollen grains moving randomly in water. He described the motion in 1827 but was unable to explain it. While pedesis takes its name from Brown, he was not the first person to describe it. The Roman poet Lucretius describes the motion of dust particles around the year 60 B.C., which he used as evidence of atoms.

DISCOVERY Brownian Movement

Discovery of Brown

Brownian movement, or motion, is a process familiar to many of us and is named after the great Scottish scientist, Robert Brown.

F **Brown's Accidental Discovery**

Whilst looking within the vacuoles of the pollen grains suspended in water under a microscope he discovered something that appeared to wobble and move almost randomly around the medium.

Now we fast forward to 1827 where Robert Brown, a British botanist, is observing a suspended pollen grain in water. While looking at this pollen grain underneath a microscope, he notices that it undergoes a type of random walk. The figure below depicts[6] the type of random, seemingly unpredictable motion, that the suspended particle underwent.

Reference books copy from each other, and so they explain Brownian Motion as the 'movement of pollen', which it isn't. The effect is actually the jiggling motion of minute particles a million times smaller than a grain of pollen, caused by the ceaseless movement of atoms. Tiny atoms could never make a pollen grain move! What these reference books claim is obviously impossible—but, if the other reference books say so, then a new one must say the same.

People rely on 'reference works' to provide information but (because they prey on each other's entries) they are riddled with marvellous mistakes. In the *Musical Guide* encyclopaedia, originally compiled in 1903 by Rupert Hughes, the final entry is 'Zzxjoanw', defined as the Māori word for a drum. Years later, someone

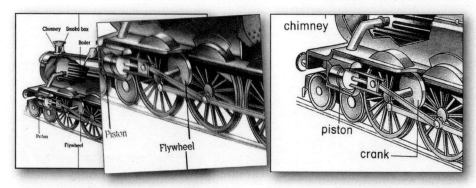

When the proofs of my *First Encyclopaedia of Science* arrived I was astonished to see that the locomotive picture had acquired a flywheel. This is a fiction—but the American editors insisted this was how it appeared in the reference books they consulted. I reckoned it might have been a mistaken interpretation of the crankshaft, but it was too late to alter the image. I insisted they changed the wording, though, from 'flywheel' to 'crank', and that was the best I could do.

pointed out that the Māoris do not have the letters J, X or Z in their alphabet ... the word and its definition were obviously fictitious. Yes, it was Hughes's joke. People have been copying it ever since. It was published in the widely used *Music Lovers' Encyclopedia* (1954), again in *Mrs Byrne's Dictionary of Unusual Words* (1974), and most recently in *You Say Tomato* (2005).

Once an incorrect entry has been dignified in print, it's very hard to correct it. I experienced this first-hand when compiling my *First Encyclopaedia of Science* for Kingfisher Books in London. When the entries were reviewed by Random House, who were publishing it in the United States, they kept coming back with corrections—not errors in the text, but instances where I was correcting the mistakes in other encyclopaedias. 'No,' I'd patiently explain, 'This is the correct explanation. It's the *others* that are wrong.' A voice on the phone would plaintively respond: 'That's not what it says in the encyclopaedia I have here.' This insistence, on copying what other reference books said, meant that we published an imaginary feature in a steam engine. Even though the book was coming out in America after it appeared in Britain, the editors were agreeable to my including a British steam locomotive (instead of an American engine, with its massive smokestack and cowcatcher). That was extremely obliging—though I was taken aback when the page proofs arrived. There, in the middle of the picture, was a non-existent object: a flywheel. Static steam engines, and vintage traction engines, do have flywheels, but steam locomotives definitely don't. It was a fiction. The editors were adamant: the other encyclopaedias they consulted all had a 'flywheel' and ours must have one too. I was equally insistent that my encyclopaedia was going to correct these annoying mistakes—but by now it was too late to change the artwork. The 'flywheel' was in the same position as the crankshaft (in fact I think a misunderstanding of the crank assembly was how this mistake originated) so we

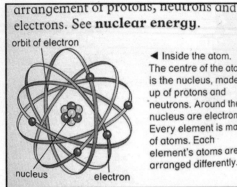

arrangement of protons, neutrons and electrons. See **nuclear energy.**

orbit of electron

nucleus   electron

◀ Inside the atom. The centre of the atom is the nucleus, made up of protons and neutrons. Around the nucleus are electrons. Every element is made of atoms. Each element's atoms are arranged differently.

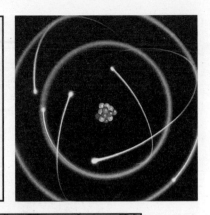

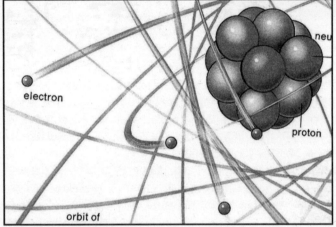

neu

electron

proton

orbit of

(*Top left*) This is how atoms are conventionally portrayed. You all learned them like this—billiard-ball electrons circulating round a nucleus made up of more billiard balls. The idea died during World War I, but it persisted ever since. (*Top right*) Because that model was so popular, I thought we could have electrons shown as centres of energy (with a hint of the old billiard ball). This impression by our graphic designer was perfect. (*Bottom*) When the proofs appeared, the American publishers had insisted on reverting to billiard balls—simply because all the other books had them. 'Don't worry,' they said, 'your balls are tiny.'

compromised. It was re-labelled 'crank' and the book was published. Sometimes corrections simply could not be made, because false facts are so widely known that they have become reality. For instance, every popular encyclopaedia showed atoms made up of billiard balls. Most still do! But that model has been out of date since Niels Bohr studied the physics of the atom—and his results were published in 1913. Put simply, he calculated that nonrelativistic quantised momentum gives an allowed velocity, momentum, and total energy in an atom of hydrogen as:

$$E_n = -\frac{m_e e^4}{32\pi^2 \epsilon_0^2 \hbar^2}\frac{1}{n^2} = -\frac{e^2}{8\pi\epsilon_0 a_0}\frac{1}{n^2} = -\frac{E_1}{n^2}$$

You see? It really is that simple. It meant the death of the billiard ball model. In light of this, I wanted my electrons to look more like energy shells, rather than the billiard balls in the existing encyclopaedias. The editors warned me that the diagram had to be compatible with all the books children used (*opposite top left*), and eventually I agreed to showing electrons as fuzzy-looking energetic specks whizzing around the nucleus (*opposite top right*). That seemed to be a sensible compromise. I was in for a shock—when the page proofs came back, the sheet on the atom was nowhere to be found. I think they'd hoped that I would just check all the pages they sent, and not notice that one was missing. But I realised at once, and asked to see it. Nothing happened. I asked again, still without a response. Eventually I stormed into the publisher's London office and insisted on seeing that page—and, to my horror, those billiard balls had returned (*opposite bottom*). 'We can't change them!' bleated the editorial staff. 'America wants the billiard balls! All the other encyclopaedias have them and we don't want to look like the odd one out.' I was dumbfounded—but the production slot was booked with the colour printers and we'd lose it if we didn't move at once. My editor offered a word of consolation. 'I am sorry the Americans wouldn't accept the change,' he said, 'But they have completely redrawn the proportions of the atom. At least you can be happy that in the whole of publishing history your balls will be the smallest.' With reluctance, I'm ashamed to say, I signed off the pages.

Similarly, encyclopaedias and textbooks all include that experiment with the candle under the glass jar. You know the one. A lit candle is floated on water under a jar and, as the candle flame burns, water rises up in the jar. When the flame goes out, the water has risen one-fifth up the jar, and they say this shows that the air is one-fifth oxygen. You've probably done the experiment yourself. I realised as a schoolboy (sorry, student) that it's a nonsense from start to finish. First, the candle flame goes out long before all the oxygen has gone. It can't keep burning until every last molecule has been consumed! Second, the jar ends up filled with smoke and carbon dioxide which occupy space of their own. Third, the smoky gas is hot. Finally, the levels of the water inside and out would have to be the same for the volume to be accurately measurable. This demonstration is riddled with mistakes. Even so, the experiment is quoted everywhere. I opened an encyclopaedia, and there it was. On a lecture tour in Malaysia I toured their national science centre, only to see a student volunteer making the same claims. Back in England I saw an alchemist demonstrating the same fictions to the public at a traditional fayre … and it's all wrong.

Now, with information mushrooming all over the Internet, it is hard to know which sites are trustworthy. People often say that they love the freedom to post anything they like online, without an editor intervening. It costs nothing. You remain completely anonymous. And you can say what you want. In fact, editors exist to stop you publishing nonsense. They are your greatest ally!

EXPERIMENT
Oxygen in the air

Adult help is advised for this experiment

In an effort to discover what it was in the air that animals needed for breathing and flames needed to burn, scientists in the 18th century conducted many experiments with air trapped in a glass turned upside down in water. This classic experiment shows a fifth of the atmosphere is oxygen.

YOU WILL NEED
● candle ● large glass jar (calibrated if possible)
● egg-cup ● shallow dish of water ● matches

1 FIX THE CANDLE into the egg-cup with a little melted wax and place in the middle of the bowl. Fill the bowl about three-quarters full with water, making sure the candle is well clear of the surface of the water. Light the candle.

2 LEAVE THE CANDLE to burn for a couple of minutes. Place the jar over the candle at a slight angle to expel some of the air, as it lowers the level of the water. Note the level of water in the jar at this stage.

3 AS THE CANDLE BURNS the water will rise up the jar to take the place of the oxygen that is being used up. When the flame goes out, it means all the oxygen is used up. You will see that the water level has now risen by about a fifth, indicating that oxygen comprised about a fifth of the original air.

This experiment appears in a standard children's encyclopaedia. It purports to show, by burning a candle in a jar, that one-fifth of the atmosphere is oxygen. At the end of the experiment the jar is filled with smoke, hot gas and water vapour (plus some remaining oxygen) so it proves nothing. But the experiment is everywhere. At a country fayre in England I watched an alchemist explain it, and at the Science Centre in Malaysia it was attracting a crowd. It is wrong from beginning to end.

Unchaperoned writing explains why there is so much misleading rubbish in circulation. Truth means nothing on the Internet, and so myths abound.

So, be careful which sources you cite—although, on the other hand, who's going to tell the difference? You may be producing garbage, but the chances are nobody will ever know. And (if you are simply copying the mistakes that everybody else has already copied) you're completely in the clear.

THE INTERNET RUMOUR MILL

It has been recognised for centuries that lies spread faster than truth. There's nothing new in that—the idea has been published for more than three centuries. Jonathan Swift wrote in *The Examiner* that 'Falsehood flies, and the truth comes limping after it' on 9 November 1710. On 5 September 1820 the *Portland Gazette* published: 'Falsehood will fly from Maine to Georgia, while truth is pulling her boots on' which may have encouraged Charles Haddon Spurgeon, the so-called Prince of Preachers to write 'A lie will go round the world while truth is pulling its boots on' in his 1859 book *Gems from Spurgeon*.

We must not imagine that the internet has a unique hold on the global spread of absurdity. That's not the point—Experts need to realise how they can harness it to spread their latest ideas. Once some non-existent 'discovery' has been posted on the internet, it can spread like wildfire. Deb Roy, Professor of Media Arts and Sciences at the Massachusetts Institute of Technology, led an investigation to show that false news spreads faster that facts. They studied 126,000 different stories that had been tweeted by 3,000,000 people more than 4,500,000 times to prove the point. You can map the way published gossip can infect the whole world and in 2015, Manlio De Domenico and his colleagues at Birmingham University published 'The Anatomy of a Scientific Rumor' (published in *Scientific Reports*, doi. org/10.1038/srep02980). It looked at the way the discovery of something which might have had similar properties to the hypothesised Higgs Boson was communicated across the world-wide web.

So you needn't worry about your research findings being rubbish. Just get them online—if they are based on sound scientific facts they won't do so well, but if they're exaggerated drivel which makes no sense to anybody of sound mind, then the world is your oyster. Internet, here we come.

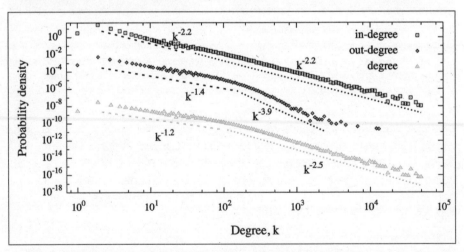

To clarify how rumours spread across the internet, De Domenico published this graph 'to show the probability density of in-degree, out-degree and total degree of nodes that tweeted about the Higgs Boson, with the corresponding distributions shifted along the y axis to put in evidence their structure, with the dashed lines shown for guidance only.' I couldn't have put it better myself.

CHAPTER 6

The Expert at Large

So now you are an Expert. You have stayed the course, weathered the storms; above all you have been selected by a time-honoured and immutable system which has granted the necessary paper qualifications—your ticket to practise—and you have a job that will sustain your financial needs.

At this stage a careful personal reassessment is necessary. There are certain codes of behaviour, dress and image consciousness that must be acquired. A failure to master them could—even at this late stage—result in a loss of standing and social status: a loss, no less, of the respect you can rightly expect from the Lay Public.

First there is the question of public acceptance. How do you seem to the outsider, the Lay Person? What do your neighbours make of you?

The Expert is always seen as a Man of the Times. He must be seen to be up-to-the-minute and *avant-garde*. Each era of Western civilisation has its own acceptable codes of unconventional conduct and it is essential that the Expert understands the correct orientation before he attempts to move in society.

The matter of dress is important. There are, at the moment, two schools of thought over this issue. Either you may be a strict conformist, and follow the dictates of modern fashion,* or you can opt to be original by conforming to an earlier mode of dress. The choice is up to you. Either mode will do; the important factor is the conformity element.

(a) The Modern Conformist

This mode consists basically of showing an affinity for the more recent conventions of attire. It is important to grow a beard wherever possible for gentlemen, and to discard a brassiere if a lady. Sandals should be worn generally. Denim jeans are favoured for both sexes, and it is imperative to avoid the convention of

---

* 'Fashion' is a special use of the term 'Fashionism' applied exclusively to mannerisms and style of attire.

wearing a necktie. A cravat or knotted scarf are possibilities if an open-necked shirt lets in too much of a draught. Sweaters are possible, leather, hide or sheepskin coats are excellent, but must be dirty. Hair should be worn *either* very long *or* close-cropped. Above all, avoid short-sightedness of the kind that blurs the boundaries between conventions: thus scarlet trouser braces ('suspenders' as the Americans have it) would not go with long hair. Do you follow? Excellent. A pair of forceps, an artery clamp, slide rules or a stopwatch are useful accessories and should be worn either in the breast pocket or dangling around the neck. However, do steer clear of convention breakers: it is imperative to avoid suede shoes, flannel trousers (even if flared) or a tied necktie. Do not carry books.

(b) The Orthodox Conformist

Here we have the principle of conformism in its other guise—conformity to earlier modes of dress. For men, suits are worn. They may be of cord, but dark well-cut polyester suits are best. Ties are always worn and may be *either* the tie of an old college/institute/club *or* should be made of the same material as the shirt. Littlewoods sell ties that are designed with a cell-wall pattern (good for biologists) and Woolworths stock Eton, it is worth noting. Shoes are shiny and modish with an offset buckle. For women the traditionalised midi is worn, though hot pants* are acceptable too.

The trend of fashion can be followed by a weekly perusal of any of the stylish Sunday newspapers (*Observer*, *Sunday Times* or the *News of the World*) and adherence is essential! The half-hearted semi-conformist will end up no more than an eccentric has-been. Remember that.

The choice of mode, whether Modern Conformist (*M*con) or Orthodox (*O*con) is up to you, but it depends to an extent on where you are going to be. The *M*con mode applies particularly to environments such as Berkeley, the London School of Economics or Essex University. The *O*con mode would mark you out better for Oxbridge, the Massachusetts Institute of Technology or industry in general. In the main, find out which mode is followed, and then do likewise.

Slavishly.

It is at this stage in your career that the use of *Quasi*-notional Fashionistic Normativity (the *qu*FN factor) comes into its own with a vengeance. Let us permit one lapse here, and allow the use of the term 'Fashionism'—it's basically wrong being an Interpretationable Abbreviationism, but once in a while we can bend the rules a little. That is what randomisation is all about, remember.

---

* If liable to goose pimples, do not be deluded by this term. It is descriptive of appearance, not function.

In social intercourse the use of Fashionistic behaviour patterns is vital. Certain expressions and terms, for instance, are Fashionistic while others are of low-Fashionism rating just now. You can monitor changes in the pattern only by studying the media—*but not by confusing Fashionism with trends in public acceptance.* It is a common mistake to confuse the two, but they may be poles apart. For example, as election results show, Western democracies (as they like to call themselves) are unlikely to be replaced by left-wing political factions. Communist candidates at elections receive tiny voting support, as a rule. Yet Communist speakers, writers and pundits are always popular and are regarded with respect.

Much the same is true of skiing, falconry and the use of mind-expanding concoctions—widespread acceptance and support is given by people who would not themselves wish to partake. In fact, participation is often painful, or actually damaging, perhaps expensive too; and so we have an important characteristic of Fashionism:

> *Fashionism implies widespread tacit support, but not acceptance or participation.*

Fashionistic elements of modern life abound. Some years ago it was the twist, modern art, pop music, and the rest; as the 1960s close it is sexual liberation, democratic socialism, skin diving and rock-n-roll. In a few years it may well become pornography, anti-drugs and anti-smoking, eugenics, UFOs, ESP.

So now we have come a considerable distance along the path towards true Expertism. You would now find yourself acceptable as a human person to your friends and neighbours, while still embodying all the best characteristics of Nonscience at its most refined. We gauge the results by speaking of your AHb rating.* By now it should be very high.

Yet for all that—and notwithstanding your excellent *qu*FN orientation on top of your paper qualifications—you are still not a fully-fledged Expert. It is now the time to consider your image. And before enumerating the principles concerned, we can list the attributes associated with Experts in the mind of Lay People. The public consider Experts to be:

wise	*infallible*
noble	*superior*
aloof	*all-knowing*
distant	

---

* An abbreviation of Acceptable Human Being. Do not confuse it with the Hb of haematology, will you?

And now all we have to do is examine the means by which this image may be perpetrated.

(A) SUPRACHARACTERISTIC SOCIOTROPISM

This is known as the SS-mode of behaviour to devotees of the Movement. It consists of adopting the correct behavioural patterns for a person of Superior Status. To a certain extent these guidelines are modified by Fashionism, but there are basic rules that *must* be followed. Do not talk to people more than necessary. Keep yourself to yourself. Nod curtly to shop assistants and other ordinary people; grin at—but do not speak to—people you recognise (or even if you don't) in the street; do not ever hang up your own coat but have it done by a menial (secretary, Lay husband, wife, etc.) in the laboratory and at home. Have all your correspondence answered in the third-person singular (first-person plural for intimates), travel by cab wherever you can.*

It is imperative to hold the head erect, walk stiffly and with measured grace, and to speak in the firmest, most low-pitched voice that you can manage without risking a froglike croak which would inadvertently precipitate ribald laughter from unwitting louts nearby. When interviewed, stare into the middle distance; when photographed, scowl. Have nothing to do with salespeople, tradesmen and suchlike and at all times see people by *appointment*.

(B) INFALLIBALISTIC OMNIPOTENTIALITY

The Expert cannot afford to be seen to make mistakes. They avoid this on the personal level by saying things in a way that can later be reinterpreted to mean whatever suits the circumstances. Thus if you confidently state that 'it's going to be fine tomorrow' you will be wrong, wrong, wrong if it rains all day. The correct formula would have been:

> 'Aha, the cirrus formations visible in the upper atmosphere suggest the imminence of a cyclonic airmass tomorrow; and we all know what *that* means!'

Whatever happens then you are bound to have been right. To take another

---

* If economic considerations make this difficult, go by public transport to near your destination and then take a cab. This applies particularly to London, where cab fares double after six miles. Though this is a closely guarded secret, Experts can use it to embarrass a rival who is ostentatiously richer. As the fare suddenly begins to escalate prodigiously the look on their face is a study. Hire cars, with a liveried driver, are very impressive if you have time to book one.

example (this time of *Implicatory Suggestionism*), rather than suggest a viewpoint that a listener might quickly reject, it is better to conclude by saying '... and do you want *that* sort of thing happening?' If the listener does, or doesn't, you can be seen to have upheld your answer well enough. A similar form of device, *Juxtapositionary Implicationism*, is used to implant an idea by an assumed cause-and-effect relationship. It is used in presenting material for publication (page 117) and takes the form:

Woman Takes Birth Pill: Dies of Thrombosis

In that particular format it is familiar to us all. No cause-and-effect has been demonstrated, indeed the chances are that none exists in that case (since more women die of this condition normally than could possibly be ascribed to oral contraceptives), but for the proponent of this viewpoint it is a handy weapon.

If we had a case where a steeplejack slipped and fell during his lunch break we could utilise the same device thus:

Habitual Tomato-Sandwich-Eater's Sudden Death

Perhaps you chose to be anti-breakfast-cereal in a research project and knew of a short-sighted relative who had been involved in a rush hour accident; you could present this:

Woman Eats Porridge: Dies of Massive Internal Injuries

Or the proponent of a rival proposition who, in old age, suffered from senile decay would provide ammunition of the form:

Diehard Ends in Mental Hospital

and so forth.

These constructions are used *to mar the efforts of antagonists*, and they are often abbreviated to the general term Marguments. These are inordinately handy devices to have up one's sleeve for the occasional difficulty that arises in public life.

The use of Marguments in debate or in writing and a careful professional guardedness about research matters ensures the correct orientation. Infallibalistic Omnipotentiality (known as IO) is a valuable aid at home, at work, in social life. 'IO, IO, as off to work I go' should be on the lips of every self-respecting Expert.

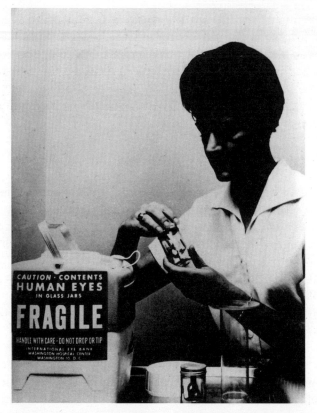

In research work, always be sure to pack your essential material in clearly labelled containers, thus avoiding confusion (at the same time encouraging awe and respect from others).

(C) ENDOFACULTATIVELY REORIENTATIONAL NORMOPROFESSIONISM

As an expert, seeming to stand alone, above the common herd, you will come to find that support from others is necessary. Your IO rating may be high, your SS mode as pronounced as can be; but without support from others—people in the right places—your efforts for success cannot get you very far.

If you are hoping for promotion you will need to have certain key figures on your side—the departmental head, for instance. Perhaps it is publicity through radio, TV or the press that you require. In this case you will need to cultivate the correct relationship with reporters or producers. This calls for the application of a cultivated ability to make social concessions to people in these roles, or Endofacultatively Reorientational Normoprofessionism (EreN, as we call it). It is important to base the relationship firmly on high-SS grounds at this stage, but bearing in mind the ideals of the individual with whom you intend to build

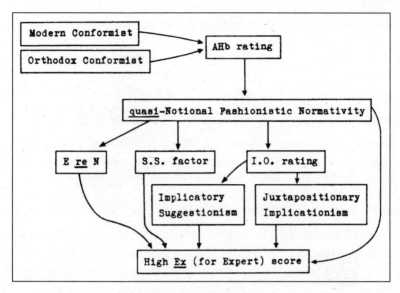

Social behaviour patterns and their interrelationships with Expertistical Proficiency.

the relationship (the *ployee*) so that your SS-orientation may be Fashionistic in the true sense. Meanwhile one's IO inclination is applied, in the form of SS-relaxation, to the ployee only at carefully regulated intervals. In this way they are flattered by the fact that you seem to be warmer and more approachable towards them than you are as a rule towards lesser mortals.

Inevitably an Identification will spring up between the pair of you, and they will be delighted by the implied compliment. At a higher level of intimacy you may even allow small flashes of humour to creep in, but beware of this device in the normal run of events, or it will look like Undue Familiarity. That would *never* do.

We saw earlier how it can be vitally important to have the right description—Professor being so much more socially acceptable than humble, undistinguished Mister, for example. In other ways it is important to have the correct name, and now is the time to consider it carefully. At the present time, odd-ball names for Experts are really very Fashionistic, but the trend could change.

As a general guide in choosing a name for yourself if your own doesn't suit, there is one piece of advice that can be given: *choose a forename that is more usually used as a surname.*

Thus, though Stan Slade isn't very exciting, Watson Slade is, John Jenkins is vastly inferior, as a Nonscience nomenclatorial designationism, to Reece Jenkins. And Al Tucker cannot match Shackleton Tucker, any more than Fred Simpson can measure up to Merridew Simpson. Some examples are shown, which illustrate the point.*

---

* Apologies in advance to anyone who actually has these names already.

Original name	Expert nomenclatorial designationism
Stan Slade	Dr Stanley Watson Slade
John Jenkins	Dr Jonathan Reece Jenkins
Al Tucker	Dr Alan Shackleton Tucker
Pete Smith	Professor Peter Townsend Smith
Fred Simpson	Frederick Merridew Simpson
Tom Gibbs	Thomas Fullerstone Gibbs

The results are always gratifying. Can't you see it now?:

'According to the results of Fullerstone Gibbs, which confirm those of Watson Slade and Merridew Simpson, the earlier theory of Reece Jenkins cannot substantiate the work of Townsend Smith and Shackleton Tucker …'

Reading that hypothetical sentence through with the original names instead shows the degree of improvement that results. If you are born with such a name, so much the better; but whether it is real or acquired the chances of a hyphen appearing are very high (Stanley Watson-Jenkins) and even if it doesn't you will always be addressed as though the name was double-barrelled.

The choice of a professional name is a matter that has, in other fields, already been widely practised. Some examples are given in the table:

Profession	Original name	Acquired professional name
Singing	Terence Nelhams	Adam Faith
Modern Cinema	Maurice Micklewhite	Michael Caine
Acting	Larushka Skikne	Laurence Harvey
Economics	Gunter Kees	William Davis
Royalty	Philip Mountbatten	The Prince Philip, Duke of Edinburgh
Ballet	Alice Marks	Alicia Markova
Publishing	Jan Hoch	Robert Maxwell
Conducting	Leo Stokes	Leopold Stokowsky
Rock Music	Clive Powell	Lance Fortune
Modern 'Pop'	Lance Fortune	Georgie Fame

It has happened in our field in the past, when it was more Fashionistic; Karl Linné acquired a 'von' as his work became better known, and before long he had become Linnaeus—which is far more Expert-like. So do consider the question of name seriously. If a change is needed, now is the time to change it: an alteration later in life will give rise to puerile comments from those too idle and uncultured to know better.

In short, the Expert who is going places needs to understand and apply the following characteristics:

Public Acceptance (AHb)—Modern Conformist (*M*con) or
Orthodox Conformist (*O*con)
Quasi-notional Fashionistic Normativity
(*qu*FN)
The Image of Expertism—Supracharacteristic Sociotropism (SS)
Infallibalistic Omnipotentiality (IO factor)
Endofacultatively Reorientational
Normoprofessionalism (E*re*N)

The mathematical expression of these interrelationships should be ingrained on the mind of every Expert who hopes to make the grade:

$$X = AHb \left[= \sqrt[2]{\frac{Mcon + Ocon}{2}} + \sqrt[2]{quFN} \right]^2 + SS + \frac{1}{EreN} \text{ (IO)}$$

Don't forget that, either. Your public image depends on it.

MEDIA BOUND

You rarely saw Experts on television in the old days. But now they're everywhere. You can see physicists, ignorant of everything in the world but their own tiny sector, pontificating like priests on topics about which they don't know anything. There are PhDs waffling on self-indulgently about trifling concerns with a perfectly honed aura of condescension, even though nothing they say has any relationship to the real world. And you can do it too! Gone is the aloof and distant manner that marked out Experts on TV when the original book appeared; now you must remember to smile. Beam at the public affably. Sit up straight. Always start by thanking the presenter for inviting you on their show. This is absurd—the presenter didn't invite you anyway (that's the job of the producer) and *they* should be thanking *you*. After all, you have given up valuable time to appear and they invited you onto the programme only because it will help them—they haven't asked you to appear because it will benefit you. Start every sentence with the word 'So ...' even when it has no grammatical significance. Always have a couple of juicy stories to quote, and if you're ever asked about an embarrassing example you know nothing about, never admit it, just say 'Obviously I cannot discuss individual cases.' You can, any time you want, but—if you utter those magic words—the questioning stops. If asked to explain some controversial topic you know nothing about, start your answer with 'Totally' then explain that, since this is outside your area of expertise, you cannot possibly comment. Don't pay attention to the questions, because they will always be inane. Instead, memorise a few great statements or anecdotes that emphasise the crucial nature of your work, and include them anyway. Remember the young Experts who appeared on television when the coronavirus epidemic was new? They knew no more about it than the public had already learned from the media, but they chatted with an aura of authority about the safety of parcels sent from China, the way the virus might be spread, how infectious it might be … without a scrap of real knowledge. Just make stuff up. Nobody is ever the wiser.

If you are being asked to explain a serious problem about which nothing constructive is being done, always say: 'We are working to find a solution to this problem day and night,' or 'we are labouring round the clock (or twenty-four seven) to find an answer'. Interviewers are always happy with that, so you can go home and relax with a coffee. If they contact you for a follow-up interview, have them told that you are too busy. Obviously you are, if it's a big problem. Should you be asked why some specific measure hasn't been taken, it is best not to admit that you can't be arsed to do it. Just reply that you cannot because of Health and Safety, and that ends the matter. The Health and Safety at Work Act 1974 was brought in to make sure that people didn't take unnecessary risks, and not to interfere with everyday activities or traditional pastimes. But the public don't realise that. They know that, if you utter the words Health and Safety, you can get away with any nonsense you like. These are the three most hated words in the English language, and they save any Expert's skin when you are caught out.

The questioning of ordinary people, like politicians, is often so aggressive that they cannot say a complete sentence before the interviewer cuts across them with an interruption. As a result, Boris Johnson, even before he became Prime Minister, refused to appear on the BBC's Today programme. Many other politicians have the same attitude. But Experts can get away with anything. They say whatever they want, they are allowed to finish every sentence, they evade every difficult question, and nobody ever queries their motives or conclusions. The interviewers are such an ignorant and uneducated lot that being interviewed on Today offers an open microphone for anything you like, if you're an Expert. Television interviews are equally easy, so long as you follow the rules.

In an interview, never say you have 'been in contact with' somebody; always say you have 'reached out to them' instead. Don't say 'from the start', it should be 'from the get-go'; rather than 'twice', always say 'two times'. This is trendy talk that you need to embrace. Then, at the end, always say again: 'Thank you so much for having me on your programme,' and give a smile at the camera (the one with the red light) to claim the loyalty of the viewer. But remember, don't talk casually to people when you have a microphone clipped to your clothes or someone will put the recording out on social media. And at the end, make sure your microphone is unplugged before you walk away. Remember these points and you can't go wrong.

Conventions of dress have changed. Fifty years ago, the necktie became unfashionable, so all trendy young men stopped wearing one. Unfortunately (in order to seem on trend) old men dropped wearing a necktie too. Youngsters look trendy and fashionable like this; old men just look scruffy. There is a critical age (about 48½ years) below which a male scientist looks overdressed with a tie, and above which they simply don't look as though they're properly dressed without one. Bow ties can be worn, of course, and those who work in microbiology laboratories need one, since a straight tie dangles in your cultures. Similar age-dependent trends apply also to women; those who wear a skirt or dress have to choose a longer look when they pass that critical age. Below that critical cut-off point you look smart and trendy; above it you look like a tart. That's also when women have to bob their hair to look on trend. It's terribly important. Faux fur is excellent for the successful woman—it has resonances of luxury while making a statement about animal conservation. Shame it's all unfashionable plastic, but the public don't realise that.

What hasn't changed is the wearing of denims; when *Nonscience* was first published I thought about jeans in a nostalgic fashion. Surely they'd die out, and become identified with ageing hippies from an earlier era? It hasn't happened. To this day jeans remain popular, and many wear them distressed or torn. It is a fashion paradox that new jeans are much cheaper than those that look old. Fashionism trends have changed; as predicted, pornography, anti-drugs and anti-smoking have become big news. Eugenics is there too (choosing the sex of a baby, selecting a listed genius as sperm or egg donor), and there are plenty of programmes on TV about UFOs and ESP. In future the centres of interest might be geothermal energy,* timber buildings, mass transport, cultured food and the death of television.

THE NAME QUESTION

This is the era of the Woke Generation, in which an awareness of racial and social disparity is paramount. You don't have to believe any of this, of course, just act

---

* This one is overdue for attention. Here we have limitless energy lying beneath our feet, though it is never seriously considered, and it is omitted from official government reports on energy futures. This is the answer to our energy crisis. Now, there's a thought.

as if you did. And you must use current terminology: calling a black or Asian colleague a 'person of colour' shows you are on trend (but never say 'coloured person' or you'll trend majorly on twitter). Similarly, when the COVID-19 epidemic began, the traditional choice of name would have been Wuhan Disease (like Hong Kong Flu). But someone would have said it was racist, so these days we use alphanumeric characters which look official and don't raise eyebrows. The public are endlessly diverted by such nuances. As an Expert, you can get away with killing people quite easily, so long as you say you have learned your lesson, but a comment that someone claims is discriminatory can have you dismissed. British Universities are inherently racist: I do not ever recall seeing a black academic at any of the innumerable dinners to which I've been invited at scores of universities over the years, and the Academic Education Statistics Agency report for 2020 shows no black person holding a senior post in any British university. Mind you, now that word is out about my book, I think that may soon change.

The question of your name remains important. When the original book was published, foreign-sounding names were the subject of discrimination. British people cannot read foreign names, and they never liked them. As Nonscience has permeated life with its sense of untouchable mystery, all that has changed. A foreign-sounding name has become an asset. Indeed, sometimes people have tried to change their Western names into something foreign and exotic. Weird spelling is equally helpful in establishing your presence (as when Danny became Danii, Connor became Konnyr, Amy morphed into Aimée, and Emily became Emmalee).

Women must carefully consider their names should they wed. The tradition was that a woman would take her husband's name on marriage; men always retained their surnames through the generations (although arbitrary, it is a curious coincidence that this parallels the transmission of the y-chromosome from father to son and thus is of value to genealogists). These days, one-third of couples aren't married. Often, the option is to have hyphenated names, retaining the surname of both parents. This can look very impressive on your office door and when your name is flashed across when you're on TV, but it presents a hopeless precedent for your children. If they all did it too, then your great-grandchildren would have 32 hyphenated names. If astrophysicist Penny Winston conventionally marries dentist Harvey Matthews, then Professor P. Winston-Matthews would look great on her letterhead! But by the time you're 70 you may have to send a greetings card to little Ben Jones-MacArthur-Gonzales-Adams-Montgomery-Graham-Wilson-Moore-Gardener-Taylor-Abdullah-Robinson-Nguyen-Nelson-قذافي-Harris-Nkabinde-Smith-Cheong-Ellis-Gunderson-Mitchell-Agarwal-Stuart-Stewart-Balakrishnan-张-Hatman-Boneville-Kameāloha-Pratt-Greenwood, which makes much less sense. People sometimes subtly edit their names so that they seem foreign and exotic (something that was frowned upon in academia back in 1971). We all know the Brontë sisters, with that alluring umlaut over the e; but few people know that the father's name was actually 'Brunty'. He thought that sounded common, so he changed it to a more up-market version. The microscope pioneer Antony Leeuwenhoek decided to become 'van Leeuwenhoek' in 1686, without any reason other than to sound more eminent. If Eric Bishop can transmute into Jamie Foxx, Mark Vincent into Vin Diesel*, and Amanda Rogers become Portia de Rossi, then the sky's the limit. Common names like 'Ford' are rubbish.

You could even try to create a name through the 'mad scientist name generator' at *https://www.fantasynamegenerators.com/mad-scientist-names.php*—anything to help you find a name people will always remember.

BECOMING GENERIC

The pinnacle of achievement for anybody is for their name to become a household word, like a biro. Its inventor was a Hungarian newspaper journalist named László József Schweiger who had a brother named György Bíró so László later adopted his brother's name (I hope you're keeping up), we will call him Bíró from now on, or Biro, since that's simpler.

Like any journalist, Biro was aware that printer's ink dries almost instantly, whereas fountain-pen ink can take several minutes to dry. He tried using printer's ink in a fountain-pen, but found (obviously) that it would not flow properly so, with his brother György, he designed a pen with a ball at the tip which, as it rotated, brought ink from a reservoir onto the surface of the paper. He exhibited it at the Budapest World Fair of 1931 and patented it in Paris in 1938. As the Nazis took power, the two men escaped to Argentina (annoyingly, this was to become the favourite destination for Nazis as WWII drew to a close) where they formed Biro Pens of Argentina and perfected the design. The pens were ordered for the RAF and in 1945 the rights were purchased by Marcel Bich, whose BIC brand since sold 100,000,000,000 (a hundred billion) pens worldwide. In the end, the name Biro became synonymous with ballpoint, though in Argentina it is known as birome. Imagine that—your own invention is a word on everybody's lips.

The best example is surely Hoover. We all hoover the carpet, and many people think it is because someone called Hoover invented the vacuum cleaner. Not so: the first domestic vacuum cleaner was invented in 1905 by a British designer, Walter Griffiths, in Birmingham. It had a filter mechanism and was powered manually by bellows. Two years later, a janitor in an Ohio department store, James Murray Spangler, patented the first electrically powered vacuum cleaner, which had a rotating brush to gather the dust and a cloth bag to collect it. Spangler could not afford to go into large-scale production, so he sold the patent in 1908 to a leather goods manufacturer he knew, William Henry Hoover. And the rest is … well, history.

---

* Vin Diesel is a movie celebrity and not, as you might think, a form of biofuel produced from grapes in France.

Just think how annoying it must be for James Dyson, who first marketed his successful bright pink cyclonic vacuum cleaner in 2003. It wouldn't sell in Europe, so he opted for Japan. People say to me how clever was Dyson to think of a cyclone as a means of gathering dust particles, but the idea is much older than James Dyson. It was dreamt up in the 1880s by John Finch, of Montrose, Pennsylvania, as a means of collecting dust in flour mills. The fine flour particles produced a fog which could suddenly explode with tremendous force, and he realised that a cyclonic dust collector could control this hazard. He patented the idea in 1885. Since then they have been popular in sawmills, oil refineries, and even in cement kilns.

The domestic dust-gathering machine was a logical extension of this tried-and-tested technology. Dyson is also famous for the 'air blade' hand dryer, which went on sale in 2006. I recall using Japanese airblade hand dryers in Tokyo in 2003, when Dyson was out there, launching his vacuum cleaner. His one original notion was the ball-barrow, a wheelbarrow with a plastic sphere instead of a wheel. That was never successful, and production was soon discontinued. More recently Dyson's have been advertising a device with a hyperdymium motor (there's no such thing; it's just a trademark). And after all that … people still hoover the carpet using a Dyson vacuum cleaner. Always remember that your name is important and—if ever you could become a household word—your fame is potentially immortal.

As an Expert, your name is your brand. Like racehorses, old cruise ships and members of the House of Lords, you have an unlimited choice of possible names you can choose, and these days it is perfectly respectable to change yours by deed poll. Why stay Penny Gubbins when you could be Penelope Winsome-Fanshawe? If you're stuck with Charlie Pratt, why not become C. Montgomery Burns? Someone must. After all, the Battenberg family switched to the Anglicised version Mountbatten, just as the German Saxe-Coburg-Gotha family thought that the House of Windsor might be more appropriate for monarchs of a place like Britain. It's all a question of marketing your brand with a name people remember. Remember that.

CHAPTER 7

The Break into Print

The yardstick of assessment for run-of-the-mill Experts in the field of Nonscience is their publications—research papers and so on. They are the only tangible proof of attainment, and the main criterion by which productivity may be judged. They are vital in job applications, and very important in the status race.

What should be put in a publication is virtually limitless. It may be a new procedure, or an old procedure re-done; an interpretation of an age-old phenomenon or a redrafting of earlier work; a new observation or an old dictum re-presented.

Papers are not usually read, and it is necessary that you remember this. It is the titles that people notice, and the abstract; so if you have something that you truly wish to have communicated to the world of Nonscience at large, then make this fact clear in the title and summary. But if this is not your aim then the use of Expertistical Obscurantism can help to discourage a closer inspection.

Thus there are two forms of publication:

- Substantive Publications are those in which some new notion is propounded—the work of Kinesics, page 190, as an instance. They are meant to be read, in other words (but see the cautionary note on the next page).
- Numericumulative Publications are those which are produced in order to 'add to one's total'. They are the variety not really meant to be read, and need not contain substantive material at all. The liberal use of obscurantism here is important, since in the Inter-Expert Status Shuffle it is sometimes found that Experts will denigrate the work of others, and publications with noticeable drawbacks are quickly seized on for this purpose. It must be emphasised that this is contrary to the principles of Expertism and Nonscience: we should try to encourage mutual assistance and not petty rivalry. But it happens still, and until the corporate consciousness of worldwide Nonscience is firmly established one must beware of it and act accordingly.

Which publications are to be preferred? The Substantive are helpful in that they encourage publicity, aid status in an individual capacity and generally tend to turn up in reference books and in the list of references in other Experts' publications. But they do require some thought and this is in itself a severe disadvantage.

Numericumulative publications are easier to do, and with their greater tendency towards obscurantism they come easier to the Expert nature. But they do not provoke as much attention and comment, and will not qualify for newspaper or TV coverage. So a liberal sprinkling of Substantive publications is helpful, but the backbone of a successful career should be Numericumulation. Plenty of quantity and a sprinkling of quality here and there—that's the basis.

EXPERTISTICAL OBSCURANTISM—A CAUTIONARY NOTE

In writing a paper or a book for a publisher in one's field, it is obviously necessary to stick to the terms of the trade, and to use Expertistical Obscurantism wherever possible. The purpose of this is to prevent one's work from being read and misinterpreted by ignorant outsiders, and perhaps to discourage an over-inquisitive interest shown by an Expert rival. By the time one has reached the higher echelons of Nonscience, as a fully qualified Expert, obscurantism is instinctive. As a reflex, almost, one would say:

> 'We have accomplished this manoeuvre by the expedient of utilising manually induced frictional heating on topographical substrate high-spots inducing a consequent oxidation/heating effect in the adjacent chemical microenvironment with inherent self-propagating proclivities inducing total combustion of the whole area, and leading inevitably to the initiation of the oxidative process in a self-sustaining mode of the contained cellular lignin core'

rather than the bald and uninspiring:

> 'we struck a match'.

But this approach (another application of which has been discussed on page 68) can be carried too far. It is easy to find a new thesis to advocate—but if the obscurantism content is too great in the summary or abstract, then no-one will be able to get the message at all. So however obscurantific the paper itself may be, it is important to bear in mind this cardinal distinction between Substantive and Numericumulative publications. Obscurantism itself is a widely used and well-documented phenomenon. To quote a *New Scientist* leading article: '... it is only a short step to the conclusion that outstanding truth can surmount great obscurity, and so to the belief that obscurity in a paper is actually a mark of virtue.'

Indeed, indeed it is.

In the writing up of technical reports the regular and frequent utilisation of state-of-the-art terminology is vital. In this respect it is helpful to refer to a nomenclature table (more colloquially referred to as a 'buzzword generator') of the kind shown in the table.

	COLUMN A	COLUMN B	COLUMN C
0	synchronised	monitored	parameters
1	total	digital	facility
2	reciprocal	responsive	concept
3	systematised	management	flexibility
4	integrated	logic	programs
5	functional	correlative	option
6	incremental	balanced	hardware
7	parallel	optical	contingency
8	compatible	third-generation	mobility
9	transitional	policy	projection

Hard-to-find gaps in reports, theses, papers, etc., are simply run off as any selected three-digit number which is then transposed in the above table. Thus we might select 582 and 437, which would generate 'functional third-generation facility' and 'integrated management contingency', both of these being terms that can usefully be employed in almost any piece of Expert writing.

Technical reports can occasionally require the use of liberal admixtures of sophisticated and specialised syntax, without which it will lose a certain aura of distinction and sophisticated savoir-faireness. To this end it is as well to memorise the Simplified Integrated Modular Prose writing system, overleaf, developed by Honeywell Computer specialists. Using this SIMP kit, up to 40,000 different sentences—each one grammatically precise and duly Expertistical—can be generated. It is important to realise that, in selecting four phrases (one from each column) it is not necessary to adhere to the alphabetical module sequence shown. Other variations are:

DACB
ADCB
BACD

The SIMP modular prose system (page 119), which by selection of a phrase from each column, can generate Nonscience-visible prose for use in reports, articles and papers. The columns can be used in several different combinations.

MODULE A	MODULE B	MODULE C	MODULE D
1. In particular,	1. a large portion of the interface coordination communication	1. must utilise and be functionally interwoven with	1. the sophisticated hardware.
2. On the other hand,	2. a constant flow of effective information	2. maximises the probability of project success and minimises the cost and time required for	2. the anticipated fourth-generation equipment.
3. However,	3. the characterisation of specific criteria	3. adds explicit performance limits to	3. the subsystem compatibility testing.
4. Similarly,	4. initiation of critical subsystem development	4. necessitates that urgent consideration be applied to	4. the structural design, based on system engineering concepts.
5. As a resultant implication,	5. the fully integrated test programme	5. requires considerable systems analysis and trade-off studies to arrive at	5. the preliminary qualification limit.
6. In this regard,	6. the product configuration baseline	6. is further compounded when taking into account	6. the evolution of specifications over a given time period.
7. Based on integral subsystem considerations,	7. any associated supporting element	7. presents extremely interesting challenges to	7. the philosophy of subsystem commonality and standardisation.
8. For example,	8. the incorporation of additional mission constraints	8. recognises the importance of other systems and the necessity for	8. the greater fight-worthiness concept.
9. Thus,	9. the independent functional principle	9. effects a significant implementation of	9. any discrete configuration mode.
10. In respect of specific goals,	10. the primary interrelationship between system and/or subsystem technologies	10. adds overriding performance constraints to	10. the total system rationale.

which may need the odd extra comma here and there. But the principle is sound and the system helpful.

In this regard, the characterisation of specific criteria recognises the importance of other systems and the necessity for the evolution of specifications over a given time period; for example, the structural design, based on system engineering concepts, adds overriding performance constraints to the primary interrelationship between system and/or subsystem technologies (SIMP A6-B3-C8-D6 A8-D4-C0-B0).

The point is clear, I trust …

THE QUESTION OF STYLE

The use of specialised language in this way is believed by Lay Persons to have some connection with 'conciseness' and 'unambiguity'. They imagine that these long-winded and precise expressions are part and parcel of an internationally agreed, clearly understood standard that is necessary to avoid confusion.

Actually, as all Experts are quick to realise, that is wide of the mark. Figures and mathematical expressions are, as we shall see on page 130, healthily randomised so that a given convention can mean heaps of different things. And in the literature of Nonscience there is a similar trend. We can only scratch the surface of this interesting study (which would take a book of its own to examine thoroughly) but its ramifications are many. One far-reaching consequence has been the effect on non-Experts and their use of specialised terms.

There was the female journalist who intended to write about β-carotene in an article about vitamins. Instead she wrote it out in full—in italics too—as *Beta carotene*: and the result? She was, of course, describing a species of turnip instead. Another writer used the term 'elephantitis' to describe a well-known tropical disease in which an area of the body undergoes a chronic enlargement. That *should* have been 'elephantiasis'. The slight alteration in wording meant that he was inadvertently describing an inflammation of elephants.

There's a world of difference between a *micrometre* and a *micrometer*, of course (in fact there isn't even a remote connection), and the use of capital letters is all-important too. Do not confuse *pediculosis* (a disease) with *Pediculus* (an insect); never mistake *penicillin* for *Penicillium* or a *penicillus*. Learn to distinguish between a 'plastic daffodil' and a 'plastic lecturer', between a 'pathological condition' and a 'pathological technologist'. Do not be surprised, on the other hand, to find *Euglena* written up in zoological textbooks as a protozoan, and in botany works as an alga (both sides like to claim such organisms as their own and it is unwise to take sides as you may provoke a riot).

Other terms splendidly conform to our concept of randomisational confusionalistic obscurantism. Take 'sterile' as a case in point: An area in which all living things have been killed (as in a nuclear test site) is said to be sterile. But so too is a living thing (such as a mouse) which produces no reproductive cells. The mouse is also said to be sterile if it does produce reproductive cells after all, but only if they are not released. The term still applies to the mouse if it is perfectly sexually developed but has no micro-organisms; and when 'sterile' is applied to milk it means that *most* micro-organisms (but not all) have been removed or killed. And finally, a tumour is said to have been sterilised when—no matter how many organisms it may contain—its further growth has been halted.

So this single term means at least half-a-dozen different things. You have to be a real Expert to tell them apart, and even then it's not easy. As a test, write down your answer to this question:

'What is a sterile mouse?'

(There are six basic answers, all different, and any combination of one or more definitions can be used to produce many more.)

But to balance up the scales a little, there are terms elsewhere in the Movement that are far more precise, indeed where a small change in only part of a word can alter its meaning entirely. Take the prefixes for 'above' and 'below'—in Greek *hyper* and *hypo* respectively, in Latin *ultra* and *infra*. Thus we can say 'ultrastructure' *and* 'infrastructure' (which both mean the same thing exactly); and we can similarly use the terms 'ultrasensitive' *and* 'hypersensitive'.

This isn't possible in some other words. For instance, we only say 'ultra-violet', and never 'hyper-violet'; yet we say 'hypothermia' and never 'infrathermia'.

In a third group of terms, something even more exciting happens. Here we can use either the Latin or the Greek term—*but they mean different things altogether!* Thus 'ultrasonic' means 'at a frequency above that which is audible' while 'hypersonic' is defined as 'faster than sound'. The latter is also 'supersonic', perceptive readers will have noticed, which gives us a whole new range of opportunities, from superviolet and superthermia to new meanings for superstructure. We can expect comics featuring Hyperman in his inframarine hypomobile any day now …

THE TITLE

Titles vary. For Substantive Papers they should clearly attract attention, whereas Numericumulative Publications would do better with a certain degree of obscurantism. Thus:

'Observations on in-vitro intranuclear aberrations induced by ultrasonic microcavitation phenomena' would do well for a paper which, in the words of Dr Paul Weiss in the journal *Science*, is 'reconfirming in yet another dozen ways what has already been super-abundantly established to everybody's satisfaction.' But a more headline-catching title, for a new theory or a personal interpretation of a more ostentatious Substantive nature might be:

'Some startling long-term dangers for the unprotected human patient undergoing experimental investigations.'

The second consideration is the desirability of finding one's work listed in index compilations, or given out by computer information retrieval services. In this respect it is important to try to include significant, *Hard Terms*.

A hard term in this sense is a solid, factual, indexable nounish word such as alcohol, urine, positron, polyethylene or extroverted. A Soft Term, on the other hand, is a word such as interesting, metabolism, certain (as in 'certain species of insects'), polymer, integrated or new. These are terms that one can often use, to be sure; especially those which advertising research has shown to be useful in catching the eye (new, unexpected, free [as in 'free radical'] etc.). But hard terms turn up in retrieval systems most frequently and assure you of a good airing in other Experts' bibliographical surveys. A title such as:

'Ionic mobility and potentiation in phenyl *p*-toluenesulphonate production from *p*-toluenesulphonate, Schotten–Baumann benzoylation and sulphonylation by benzenesulphonyl chloride replacement of benzoyl chloride utilising light petroleum recrystallisation as a chloride raw material purifying agency'

would get a dozen-and-a-half cross references in any respectable index. But the same kind of work presented as:

'Observations on certain aromatic amine reactions involving caustic soda'

would hardly get you anywhere.

Note that we have two examples each of Substantive and Numericumulative titles in the above selection. One assumes it is *quite* clear which is which.

INTRODUCTION

This is where the fun starts. It is essential to observe the basic rules of play:

(a) for SUBSTANTIVE papers a clear explanatory sentence should lead into the argument thus:

'It has been widely postulated[1,2,3,4,5,6] that ultrasonic vibrations as used in diagnosis may have long-lasting damaging effects on human subjects. It was decided to investigate certain aspects of this disturbing suggestion' (Note: the references [1,2,3,4,5,6] are discussed further in the section on page 133). The rest of the paper, until we come to the summary or conclusions, may be as obscurantific as you like. It should anyway, in a paper like that, be Highly Mathematicalised.

(b) for NUMERICUMULATIVE papers it is as well to put the reader off from the start:

'Intra-nuclear mechanical disruptive microcavitation phenomena associated with diagnostic and experimental uses of ultrasound in the medical/ biological disciplines with its concomitant potential proclivity for heredito-genetic traumative consequences ...' (which has, so far, taken us just half-way through the first sentence of the paragraph above) and so forth, is infinitely more Nonscience-viable.

In general, an Introduction should set the scene for your work. But it would be quite impracticable simply to write a summary of the work that had gone before, and then outline what you had hoped to do; as we all know, Nonscience does not work like that. Ideas change, results have to be manipulated, figures go wrong, experiments fail; and rather than waste the reader's time with this form of tittle-tattle it's better by far to adopt a more succinct approach. State the *relevant* background (i.e., that which supports your own research) then outline—as though you were predicting what you hoped would happen—the investigations you propose to carry out. Give it an immediacy of style and a certain unpredictability if you can, so that the reader will be greatly surprised and gratified when it works out all right in the end. Salesmanship here should be the guideline, without it you and your work will never catch anyone's eye.

METHODS

Here is a section of your paper which simply must sound right. Quote temperatures, magnifications, atmospheric pressure, ambient humidities; remember to include solution strengths (as decimal normalities, perhaps), makers' names on apparatus and so forth. Make it look comprehensive; show you've done your homework. But do not give away too much here, or someone will likely copy your technique and publish it more widely, thereby taking the credit. This, again, is somewhat unlikely—but all is fair in the status scramble (and it is a ploy you can perhaps bear in mind for your own use, if times get hard).

Remember to make it clear whose work this is, by the way, as in this example culled from an English journal:

'I have at this moment in preparation a paper advancing a mathematical hypothesis I have arrived at, which appears satisfactorily to explain this result of mine.'

There are many ways in which this sentence could have been written, for example—

'I have at this moment a paper in preparation, advancing a mathematical explanation which fits these facts.'

But this is so weak, so flabby, so unassertive alongside the Better Style shown above. 'I have ... I have arrived at ... of mine,' by its repetition, reminds the casual reader just who has done it all. And that *is* important.

OBSERVATIONS

Be careful not to reveal too much here.

RESULTS

Here you expand a great deal on all that went before. In essence the best way to make the section flow methodically is to refer back to your introduction and merely go through, point by point, showing how the measurements/observations/data you have accumulated fit neatly into that ostentatiously objective prediction. But do not go into details yet! Save the actual 'hard-sell', the hammering home of the point for the ...

DISCUSSION

... which most Experts find hard, if not impossible, to write. Some manage by simply rewriting, in slightly different words, the Introduction and Results sections in alternate sentences, with enough syntactical re-jigging to make the resemblance less than immediately obvious. Others do not discuss anything at all, but prefer to indicate how difficult and delicate the work really was. Still others skip the Discussion altogether, but put this section heading in half-way down the Results section (which serves very well in many cases). But the most important aspect to remember is that someone—no matter how unlikely this may be—could well be reading the paper and looking for clues to a better way of doing it. Or a rival might be scanning for mistakes in your argument.

So anticipate this (see also page 118). If you feel your results may be hopelessly wrong, add 'correct to within one order of magnitude.' If you realise that you missed out some point altogether (so easy to do) then explain that this factor has been 'eliminated in the calculations'—not eliminated *from*, which gives quite the wrong impression.

Be prepared for trouble and comeback, in other words.

ILLUSTRATIONS

If Nonscience Specialisation is to be preserved, it is basically important for illustrations to be recognisable only by those who are involved in similar work. Photographs of micro-organisms, for instance, need not be clear or even

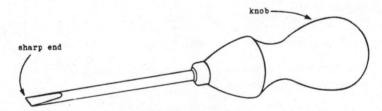

The Lay approach to a diagram.

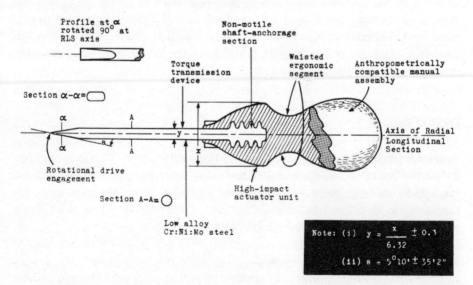

Correctly orientated Nonscience-viable diagram form.

A revealing impression of the appearance of micro-organisms under the microscope. This kind of representational copying is to be deprecated.

recognisable as Experts in the field will immediately spot what they are meant to be. In certain circumstances it is permissible to utilise a negative, rather than a conventional positive, print; this makes it virtually impossible for anyone not in your speciality to know what on earth it is meant to represent. If a line drawing is to be used in such a case instead, it should be of the type shown in order to preserve the mystique. The correct kind of drawing can be found in any standard textbook, and it is important to become *au fait* with these conventions, as they are quite unlike any organisms that actually exist.

The same comments apply to diagrams. It is clearly foolish to label a knob a knob when it is, from a functional point of view, a Process Initiation Actuator Assembly. Details of cut-away features should be included, ancillary apparatus should be drawn in too, so giving a useful feeling for the true complexity of the subject in hand. In featuring photographs of equipment arrays the same comments apply. They should not be recognisable by all and sundry, there is no merit in making the experiment look simple and lifeless by the use of uncluttered photography.

Tables of figures are a helpful way of illustrating a paper, and they too should be given considerable thought. Use plenty of columns with a full allocation of sub-divisions.

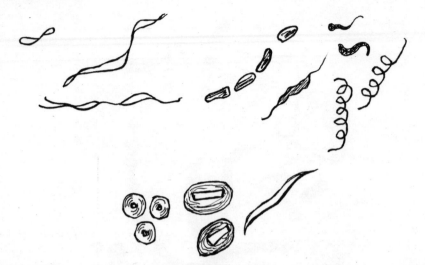

Representations of some 'common bacteria' from textbooks in current use. Note that they have a most characteristic appearance, which is—and do note this point, Experts—quite unlike any organisms that actually exist. It is this use of generally accepted conventions of Nonscience Orthodoxy that prevents us from giving away too many 'trade secrets' to members of the Lay Public who would probably not know what to do with such information anyway. The accompanying legend should state: *very highly magnified* in a duly eye-catching typeface.

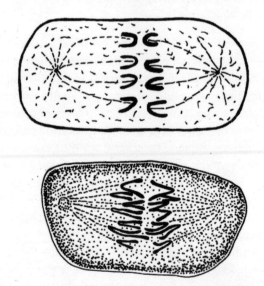

(*Top*) The dividing cell viewed with the student of Nonscience in mind. Note the adherence to the principles of movement. An interesting item of relevance to the Randomisationary qualities of Expertism is the alignment of the U-shaped chromosomes: they point the wrong way. This convention is rigidly upheld in many textbooks the world over. (*Bottom*) The same specimen drawn with the age-old pedantry of 'science'—note the lack of scope for inventiveness and originality of interpretation inherent in such sterile copyist format.

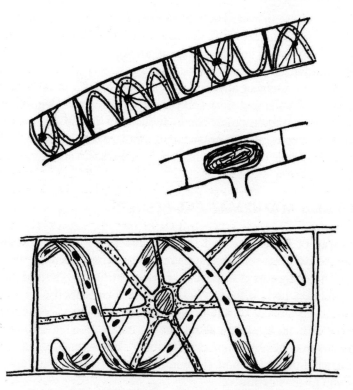

Diagrams of the widely studied alga *Spirogyra*. Note that these textbook versions reveal little of the true appearance of the algal cells themselves, thus leaving plenty of room for the Expert to interweave their own interpretation with the basic pictorial representation. It is, we must remember, patent folly to give away too much at one time, and these carefully Nonscience-orientated diagrams delicately obscure many key points.

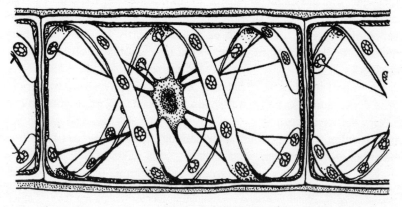

The actual microscopic appearance of *Spirogyra*. With illustrations as hollowly self-evident and clear-cut as this, what is there for the Expert to unravel and explain? Avoid such pedantry at all costs.

And then there are graphs. These have several clear-cut advantages: they are pleasant to the eye, and so break up a paper in an interesting way; they spin out a short dissertation until it covers a lot of space; they convey a great feeling of communication and understanding to the casual reader; and finally (perhaps most important of all) they can be most helpful in proving what it is desired to prove. By the use of specially drawn graph paper (logarithmic for instance) apparently random points can be induced to line up as desired. And the plotting of even a conventional graph along x and y axes chosen to fit the spread of your data can enable one to see quickly which points do not fit your prediction.

FIGURES AND MATHEMATICAL RESULTS

One easy way of getting around this problem is the use of graphs already outlined in the previous section, but there comes a time when the direct, unvarnished quoting of figures is important. Obscurantific Enumeration is part and parcel of your task here, if the job's going to be done properly.

An essential mainstay of this is the question of Degree of Accuracy. This matter is not difficult to understand.

Clearly you may be absolutely certain about certain quantities—the number of grains of sand in a salt spoon, the number of chairs in a small room, etc. The accuracy you have is definitive.

But there are other quantities about which it is hard to be so accurate—the number of sand grains on a beach, or the number of chairs in the average dining room. Here a guess (generally referred to as an estimate) is necessary, since it would be impossible ever to find the exact answer. Again, you could measure the length of a small metallic bar with a considerable degree of accuracy, but not the exact height of a man. Where is his 'top'? And what happens as he breathes in or stiffens slightly? This will certainly make a difference of a quarter of an inch or so, and clearly to talk about height to the nearest millionth of an inch would be impracticable under normal circumstances.

But the Expert just has to be precise, and so they have often to overlook this matter of degrees and settle for a firm amount. And they are helped in this task by the very nature of figures.

Small numbers, about which you can be precise, are easily defined by decimal notation.* Thus half of a cherry genoa cake (which could never be *exactly* half, for obvious reasons) would be written as 0.5—it might be 0.49, but 0.5 is near enough. If it is half a metallic bar you are talking about then you

---

* There are some figures—1/3, 2/3, etc.—which cannot be adequately expressed in this form. But don't let that worry you.

would write 0.5000, or 0.5001 or 0.4999; figures with extra noughts or decimal numbers behind them to show just how carefully you have been measuring. And the suggestion (if 0.5 is all you commit yourself to) is that you need not be too accurate if you don't want to.

Now, with larger figures this doesn't happen. Mathematicians have ingeniously left a hole in our notational conventions. Thus we say that 'there were 5,000 people in the crowd.' What does it mean? That there were five thousand, so that if one extra late arrival turned up it would become 5,001 exactly? Of course not—it means that it looked more than 3,500 but not as many as 7,000; it is a guess (or estimate) in other words.

But the use of all those noughts implies a degree of accuracy *even if you haven't any means of measuring it that well*. But you simply have to use the technique properly.

What we do is this. We take any figure that is representative of the sum we have in mind; and we guess what it might be. Thus in round figures we might have the wage of a worker in a Yugoslav factory as 5,000 dinars; or an insect flying at 10 metres per second.

Now, these approximations are simply not impressive enough for publishing. They lack authenticity and authority. They require redefining in terms more fitting for Nonscience.

The answer is an elementary one: round figures are simply converted into other units where the implied degree of accuracy is greater. Thus we convert the Yugoslavian wage packet into sterling and the fly's velocity to feet per second and at once the figures spring to life: the weekly wage of the skilled worker has become £16·46½ and the insect was flying at 32.808 ft/sec. Clearly this is a vast improvement. It suggests that if the wage packet had been a penny out, you would have noticed at once; and even if the merest tailwind had given that fly a little boost it would have swung your instruments right across the dial, so accurate were your measuring techniques.

The question of orders of magnitude is another interesting one, with obvious applications to the manipulation and interpretation of results. It may be that the obnoxious know-all reader of your work may suspect that a given figure is quoted at a far greater Degree of Accuracy than could ever be justified. If there is a real risk of this happening, then it is always possible to use a disclaimer thus 'These results are correct within an order of magnitude.'

What does this mean in practice? Suppose you have a figure in mind, say 500; but the thought occurs that you could be way out. The correct figure could be 10 times greater—i.e., 5,000—or ten times smaller, only 50. See how that would look in print: 'The figure is probably in the region of 500. It could be as high as 5,000, or again only 50; to be frank I not too sure about it.' How do you suppose that would look? You are right—dreadful. It is altogether too revealing. It lacks authenticity and firmness.

The thing to do is:

(a) Round off the figure to a quotable amount, say 493·5.
(b) Use the correct terminology outlined above.

Thus: 'Bearing in mind the influence of the many interrelated parameters applicable to the calculations involved we arrive at 493·5 as the working mean (correct to within an order of magnitude).'

That means virtually the same, but it better fits the prerequisites of Nonscience orthodoxy.

Taking an average is another way of expressing round figures in a suitably impressive form. The result is invariably impressive, and is referred to as a *mean figure*, since 'average' is altogether too chatty. But see how effective the gambit is in practice. Thus: state that the average American makes love a couple of times a week, and that is bar talk, nothing more. But study the subject as an Expert, take readings and figures (as Kinsey did), present them in a sufficiently matter-of-fact manner with dignity and authority, and the matter resolves itself as a mean intercourse rate of 2·64283, which is *far* more satisfying.

As the Expert must realise, there are averages and averages. You can take the half-way figure between greatest and smallest amounts; totalise the figures and then divide by the number of figures; or do it the hard way with a statistical average formula. Here are some examples:

(a) the *arithmetic* mean;

$$M = \frac{1}{n} \sum_{j=1}^{n} x_j$$

(b) the *weighted* arithmetic mean;

$$M_w = \frac{1}{\sum\limits_{j=1}^{n} w_j} \sum_{j=1}^{n} (w_j x_j)$$

(c) the *geometric* mean; and

$$\log G = \frac{1}{n} \sum_{j=1}^{n} \log x_j$$

(d) the *harmonic* mean.

$$\frac{1}{H} = \frac{1}{n} \sum_{j=1}^{n} \frac{1}{x_j}$$

It is often worth quoting the formula, especially if it is set in type with that flourish which the best technical publishers know so well, for its aesthetic merits as much as anything else.

Formulae always look good in published work. One good example of the kind I have in mind is:

$$P = P_0\left[\left(0{\cdot}10 + 0{\cdot}55\,\frac{L}{L_0} + 0{\cdot}35\,\frac{M}{M_0}\right) - 0{\cdot}0612 \right]$$

This formula was used by Rolls-Royce in their calculation of profit from the RB211 engine they were building for Lockheed. It is not recommended for general Expert use just at the moment.

SUMMARY

This, basically, should tie up the whole thing. A summary is usually not as long as the rest of the paper, but it might be longer than any of the component sections (Methods, Results, etc.). For a substantive paper it should be written plainly, so that the thesis or result you are advocating is quite clear to all, as opposed to the form of presentation in the paper itself, which can usefully be obscurantific. If the paper is intended to be numericumulative, the summary should be long and complex.

ACKNOWLEDGEMENTS

Be sure to list all those who have assisted, but do not allow this to detract from your own primacy in the work. An example of how not to do this appeared in the *Journal of Biological Chemistry* recently:

'I wish to thank Dr Lester who not only suggested most of the experiments herein but greatly helped in their interpretation, and A.S. Bottorf and B. Fravel for their excellent assistance in performing these experiments.'

This may to some give the suggestion that the author did nothing himself. Don't risk that. A better form would have been:

'I am grateful to my colleague Dr Lester for his suggestions, and to Messrs Bottorf and Fravel for experimental assistance.'

Put in anyone famous in the field with whom you may have discussed the work, no matter how briefly; in fact even a passing 'hello' at a conference will do, since anyone exalted enough will not mind being quoted and will certainly not remember not having met you. But do not put the responsibility for your view on their reputation:

'The author is grateful for discussions of certain aspects of his work with many colleagues, including Lord Colloid and Dr E. Coli of the Royal Society.'

They will be flattered, and you gain enormously into the bargain.

REFERENCES

This is a tricky one. The list of references for a published paper should appear according to the editor's preference (which they will tell you about if you write and ask) and there are many variations on the theme. A bibliography must be very, very long. Many Experts simply collate the tables of references at the end of earlier papers and then add a few of their own, making their list inevitably longer than anyone else's.

It is very important to include the right papers in the list. If your submission is going to be sent to Lord Colloid for an opinion, and he hates Dr Coli with a rare degree of bitterness due to some long-standing personal feud, then it is not going to do well for you if you refer to several of Ted Coli's papers but leave Arthur Colloid out of the list altogether. The reverse would apply, of course, if Ted was going to review your paper rather than Arthur. So try to find out who does the refereeing if you can and make sure they get a generous mention.

I recall a paper I once submitted in which I did not refer to the work of an Expert whose closest friend was apparently the referee—'this work really ought to be mentioned,' they said. But by the time we had added the point and resubmitted the draft, a new editor for the journal had been appointed and they wrote 'this is rather superfluous to the main argument'. And a British microbiologist working with some new organisms found another variation on the scene when she submitted a draft to two journals. One of them rejected the paper as being 'too superficial and far too preliminary' while the other said it was 'unoriginal and only described work already well known.' Now with the randomisation element in Nonscience this is healthy enough—but with a little care, as we have explained, this drawback can be neatly circumvented.

Certainly it is helpful to have the text of the paper liberally dotted with references (page 124). It is far more important, in writing a publication, to show you have incorporated all the previously published ideas rather than merely print something new yourself. A basic rule-of-thumb is this: A PAPER WITH LOTS OF NEW IDEAS BUT NO BIBLIOGRAPHY—FAILURE. A PAPER WITHOUT ORIGINAL CONTENT BUT A FULLY COMPREHENSIVE BIBLIOGRAPHY—SUCCESS. That never, but never, fails.

BOOKS

Writing a book is very good for kudos and prestige in academic circles: it may not be read by many people, and it does not really matter if it is never sold in any quantity either. The fact remains that you have written a book, and that alone is good enough to impress anyone. In the academic sphere there are two types of book you can write (volumes of a popular nature for more general consumption are considered later).

Specialist works

These are volumes about your work of a strictly esoteric nature. They are bought widely by libraries, etc., and sell at a very high price. No-one reads them (if the book has been sufficiently esotericised no-one should *be able to* read them) and so almost anything can go in them: personal beliefs, private banner-waving, 'in' comment and so forth. A book of this kind should be an opportunity to get everything off your chest in specialist language, and all the diagrams, graphs, results and notes you have ever put together can be simply collated for publication. It is sheer folly to make such a book easily readable or, for that matter, comprehensive; but it should be long. Publishers don't offer much in the way of an advance for a work like this, perhaps £50 (£750 in 2020) and the royalty they give may not be too generous either. But you can always boost the figure offered by a few percent (from 10% to 12½ %, for example) by offering to forgo the advance. Or if your financial needs are pressing you can sometimes get a larger advance if you stick at it, but then you may expect a lower royalty. Publishers in 1971 usually pay around 60p (roughly £10 in 2020) at the very most for the printing and binding of each hard-back copy, so they sell for perhaps five guineas (about £80 in 2020). Don't worry about a high selling price, either. Since you will be getting a percentage of the retail price the more it sells for, the more you'll get. Remember, to the innocent reader, a book must be expensive or it is clearly not worth buying at all.

Textbooks for teaching

Here you simply have to fit your writing pattern to the market in the best traditions of commercialism. The cardinal rules are as follows:

• FIND A MARKET, which means, quite simply, look around for a field in which there is a new subject or a new series of courses starting, and determine to write a book for that market. In recent years we have seen subjects such as science and society, ecology and conservation, etc., coming into this category.

- STICK TO THE SYLLABUS. Once you have found a new course (or an old one for which there are not too many recently published textbooks) get hold of a syllabus and stick to it rigidly. Do not attempt to question it or alter it around, for that is clearly beyond your scope (and if everyone starts doing that then where will Nonscience end up, for heaven's sake?), but use it as the skeleton for your book. If a publisher asks for a synopsis, it is the syllabus (with some comments of your own) that they really want to see. They can then be sure the book will sell.

- WRITE FOR THE TEACHER, and not for the taught. Many authors have fallen into this old trap in the past, but a moment's reflection shows how necessary it is. The book is ordered by the lecturer or teacher, and not by the students and it is those you need to please. It must be written at their level and cover their subject in the way they understand best. It is inevitable that this tendency makes the book a little harder for the student to understand, but that is part of our educational system. If it was otherwise we wouldn't need lecturers at all, would we! And dangerous talk like that never gets us anywhere.

- FINALLY remember your earlier rules (page 103) in order to make the book fit the vogue of the time—and there you are. The publisher's cost per paperback copy in 1971 will be around 25p (about £4 in 2020) with the book selling for £2 (£30); and with classes of 30 students being persuaded to buy the book in colleges all over the place you can soon accumulate very substantial sales. It's a very comforting feeling to think that all over Britain and Europe too, even the United States if you're lucky, students are grappling with their copy of Burton, or Jones, or Effington; and it's *you*! That's real power, that is.

- TITLES for books are obviously important. They have to reflect the image of the writer and the subject. An example is to be found in one of Blackwell's scientific works. It was originally entitled by the writer *Science at Home*, since that is what the book was all about. However, this clearly would never have appealed to the Expert eye, and the publishers knew this. So a new title was thought up: *Science of Home Economics and Institutional Management*, and the book since then has sold very well.

Now it has an identity, it utilises the correct terminology ('jargon') and neatly corresponds to the fashionable concepts of the people who are expected to buy the book. Similar comments could apply to many other examples.

At a higher level there are always the opportunities for consultant editorships with journals, who like to feature a long list of notable Experts in their advisory boards. Knowing the right people is the secret here; there is certainly no need for you to know overmuch about the subject at hand. And one can always become a referee too, of course. That gives considerable power and a sense of prestige to boot. But as a contributing editor to encyclopaedias and suchlike a very satisfying additional income can be derived, and the standards of such works of reference are generally so very low these days that almost anything goes. This should not encourage one to thoughts of slovenliness, nor should it suggest for even a moment that a hurried or ill-considered contribution is enough. Far from it—the present state of affairs means that your own work stands a good chance of shining, far better than if the standard generally was very high. I trust the point is taken?

Always bear in mind that educational publishing (as it is called) is a very profitable field for the publisher, and that virtually any book on almost any subject will sell enough to prove worthwhile.

Steer clear of agents, who will tend to suggest that they are entitled to a commission on anything you do, in exchange for the dubious privilege of 'representing your interests'. That usually means that a publisher's offer direct to

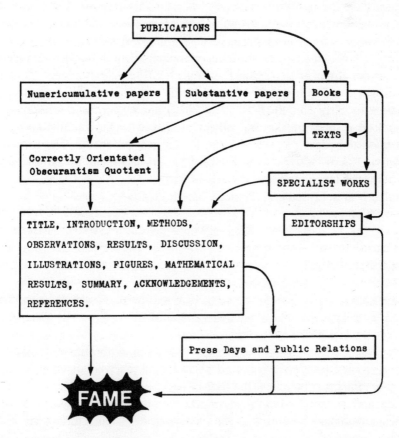

Note how this flowchart illustrates the manifold ramifications of publishing and public relations in the quest for fame and influence. Do not try to short-cut the system; it is part of an establishment whose inactivity is the bedrock of your power. Guard it well.

you is still regarded (by the agent) as something they ought to 'handle' for you. So do not trust them. Remember:

(a) agents are Experts in their own way
(b) but they have none of the training or background.

Like others in that category, ignore them.

If you have already a few titles under your belt, then you can even expect to make money out of authorship—and big money too, if you're lucky. You can now demand from a publisher an advance of at least £100 (£1,500) and royalties in the 10% range—perhaps even rising to 15% if you drive a hard bargain. The kind of book you write depends on Fashionism more than anything, of course, but textbooks always boil down to the same thing: find a syllabus and write it all out clearly, and there you are.

Guard the work carefully, as soon as you are in this category of an Established Author, and remember that book publishers are *no more than wood-pulp sales-men who make a large profit because of the elaborate marketing arrangements involved.* Do not let anyone do you out of your rights in the work—retain copy-right, for one thing.

If all goes well you can now hope to make big money. One author I know of makes an extra £5,000 (£75,000 in 2020) per annum on a school biology textbook and there are several others around who have a regular royalty income—and believe me, this is serious!—of over twice that. There are other learned works of reference which clearly deserve to make a vast fortune for their authors,* but most do not do as well, sadly.

But keep at it. From Perseverance Cometh Prestige.

———————— —w— ————————

PAPERS IN PRINT

Looking back at the 1971 edition, we can see that getting into print in the 2020s is even more important for every Expert. One of the commonest avenues is to have your work published in the proceedings of a conference. Everybody has their lecture produced as a chapter within these books, so there is little effort needed—the lecture has already been compiled, so there isn't much you need to do. As we have seen, the other people at the conference knew what you were going to say anyway, and everyone knows that nobody is going to give away any secrets or reveal anything new, because they want to keep their research for themselves and

---

* It is only the writer's innate retiring modesty that forbids him suggesting that the masterly tome you are fortunate enough to hold in your hand should rightly come into this category.

the mortgage still has to be paid. The main advantages of attending a conference are to have a holiday paid for by your department, to network socially, and to pick up scraps of gossip. When drunk, researchers will reveal their innermost secrets. Indeed, I think that we should abandon publishing the proceedings of a meeting. It would be far more revealing to publish the proceedings of what was said in the bar.

The advice in the original *Nonscience* book (namely, that it's the number of papers you publish, never their quality) is truer now than it was back in 1971. Every time a distinguished researcher is being introduced, people will say '… and author of over 50 published papers!' without reference to what the papers contained. I know a professor who has published precisely the same text more than once, each time with a different title. You'll have seen that back in 1970, it was recommended that titles should either be clear, or they should be obscure. These days a paper is reckoned to stand its best chance with a title about 10–12 words long, but long titles remain in vogue. In the *Journal of the American College of Cardiology* you'll find 'Cost-effectiveness of Transcatheter Aortic Valve Replacement Compared with Surgical Aortic Valve Replacement in High-risk Patients with Severe Aortic Stenosis: Results of the PARTNER (Placement of Aortic Transcatheter Valves) Trial (Cohort A)' though I also like 'Joint quantum chemical and polarizable molecular mechanics investigation of formate complexes with penta- and hexahydrated Zn^{2+}: Comparison between energetics of model bidentate, monodentate, and through-water Zn^{2+} binding modes and evaluation of nonadditivity effects' from *Computational Chemistry* and I also have a soft spot for 'Three-Dimensional Quantitative Structure–Property Relationship (3D-QSPR) Models for Prediction of Thermodynamic Properties of Polychlorinated Biphenyls (PCBs): Enthalpies of Fusion and Their Application to Estimates of Enthalpies of Sublimation and Aqueous Solubilities' that was published in *Industrial & Engineering Chemistry Research*. My favourite word from those titles is 'Nonadditivity' which sounds like a word made up especially for this book (but it isn't; there are over 200,000 websites mentioning it).

Personally, I think long and obscure titles are a stupid idea.

However, there is a now third option—to have a title that's funny. There is 'Fifty Ways to Love Your Lever' in *Cell Review*; a research paper on narcissism in the *Journal of Personality and Social Psychology* entitled 'You Probably Think This Paper Is About You'; one on cancer called 'miR miR on the wall, who's the most malignant medulloblastoma miR of them all?' that was published in *Neurological Oncology*; a study of piracy in the *Journal of Political Economy* that the authors called 'An-*arrgh*-chy'; and another called 'Fantastic yeasts and where to find them' in *Current Opinion in Microbiology*. The field is wide open for more.

You may find it useful to have some co-authors. Usually, the head of your department will want to have their name put first. This applies even if they have done nothing that affects you, they don't know anything about your work, and probably didn't really approve of it anyway. You need promotion—so humour the boss. You can add co-workers if these people have assisted in your project and you may want favours from them in future. Indeed, some people have added so many co-authors it is hard to keep track. Papers with over 1,000 authors are becoming increasingly common (this is not a joke, made up to amuse the reader, it's serious). This is known as *hyperauthorship*, a new term dreamt up by Blaise Cronin at Indiana University, Bloomington. Between 2009 and 2013, 573 papers with over 1,000 authors each were formally published. Between 2013 and 2018, that total rose to 1,315. The paper with the largest number was published in *Nature* in 2015, with 5,154 authors. It was a paper on the Higgs Boson from the good people working on the Large Hadron Collider. Yes, it seems ridiculous to ordinary Members of the Lay Public—but in Nonscience this is a triumph. The names of authors was always significant; now what matters is the number you can list.

WRITE A WRONG

Once your masterpiece is written, it is normal for you to recommend referees. Do make sure that you write something that reflects the views of your peers, ensure they are flatteringly mentioned in the text, and then have your paper reviewed by them. That system cannot fail. Problems arise when a second referee who you don't know is asked to comment. The current trend is for these people to glance at your bibliography and, if they're not mentioned, to be rude about your submission and reject it. I can imagine some sceptic thinking: 'Here goes Ford, exaggerating again, just to make an amusing point; that can't be true!' But it is. One recent referee's comment said: 'The phrases I have so far avoided using in this review are "lipstick on a pig" and "bullshit baffles brains".' Another was: 'The author's last name sounds Spanish. I didn't read the manuscript because I'm sure it's full of bad English.' This is a genuine difficulty, indeed a Facebook page called 'Reviewer 2 must be Stopped!' was set up recently. It soon had 2,500 subscribers which rose to 30,000 within a month.

Referees can raise interesting questions; for instance, I once had a colleague who was told his paper was rather too long for their journal, so he reformatted it with a slightly closer spacing and a reduced font size and sent it in again. It was accepted. Perhaps you cannot actually write anything cogent. If so, then simply throw in something lifted from somewhere else. People plagiarise other work all the time. You simply cut and paste copy from another publication and use it yourself. In Asia it is normal; China publishes half a million papers a year, one-third of them stolen from someone else. In India they now have sanctions against plagiarists, defined as 'the practice of taking someone else's work or idea and passing them off as one's own.' If 10% is copied there are no sanctions at all. If 10% to

> "The author's status as a trans person has distorted his view of sex beyond the biological reality."
>
> "Despite being a woman, the PI was trained by several leading men in the field and is thus likely adequately prepared to lead the proposed research."
>
> "The first author is a woman. She should be in the kitchen, not writing papers"
>
> "The author's last name sounds Spanish. I didn't read the manuscript because I'm sure it's full of bad English"
>
> "This paper is, simply, manure"
>
> "The authors study design setback the advancement of the field by 20 years"
>
> "I said that I'd never again cite or review a paper written by [XX] so it pains me to learn that this is one of their students. God help them."
>
> "What the authors have done is an insult to science"
>
> "You should look closely at a career outside of science."
>
> "[X] tried this in the 1990s and failed and he was more creative than you".
>
> "The phrases I have so far avoided using in this review are, "lipstick on a pig", and "bullshit baffles brains.""
>
> "In short, this piece of research bears all the hallmarks of some bright people who saw an opportunity in a currently hot field of research, and thought they would jump in because, after all, how hard could it be? I have scanned the resumes of every one of the authors, and have come to the conclusion that they are indeed very bright people who could have used some good advice before starting this. The passage of this manuscript would have been much easier, and I would not have had to work so hard"
>
> "This is obviously written by a group from a lower standardized institution based on the quality of work."
>
> "This person works for an NGO, you shouldn't believe anything they say."

These recent reviewers' comments have been posted on Facebook. People can be very rude about you if your research breaks boundaries. The answer is to ensure that you (a) flatter them, (b) quote their research, and (c) enrol them as co-authors or list their names in your submission. Of course, if they are second referees (and anonymous) then you're stuck.

40% of a paper has been plagiarised students have to rewrite their paper. If the extent is 50% then the student is suspended for a year and their supervisor actually loses their annual pay raise. If 60% or more has been plagiarised, then the student is thrown off their course, and their mentor has to lose two years of pay increases. I know of no other country where they take such a stern attitude. In the Western world it's an everyday occurrence. There is a paper in the *Journal of Women and Social Work* entitled 'Our Struggle Is My Struggle: Solidarity Feminism as an Intersectional Reply to Neoliberal and Choice Feminism' which lifted chunks of Adolf Hitler's *Mein Kampf* after changing a few words. All you have to do is to be careful what you copy, and where you paste it.

Now that universities are keen to make sure everybody gets through their exams, you are never going to be thrown off the course, no matter what you get up to. One of my colleagues was a microscopist at Cambridge University and one day she happened to call into the administration office only to spot a signature like hers on a document she'd never seen before. It was a forgery—one of the students she was supervising had faked a letter of approval. Forgery is a crime; until recently, that would've meant instant rustication. Not anymore: the student was merely told she'd not now have a first-class honours degree. It would be something lower. You see? These days you can get away with anything.

One avenue worth exploring is the 'meta-analysis'. This is an examination of the results everybody else has already obtained, so you don't have to do any investigative work of any kind. You can do this in a notebook while watching television, and it rarely takes more than a weekend. You will then be credited with concluding how many people eat chocolate in bed, die while playing snooker, fantasise about sex, or drink cocoa. The results are all there in the published literature, ready for you to harvest at the click of a mouse. Meta-analyses attract huge grants, but actually cost nothing. It's money for old rope, and means you aren't actually plagiarising anybody.

My research is regularly plagiarised, and I dine out on that fact all around the world. It causes me no problems at all. If somebody pinches your ideas, it shows how much they value your work, and also proves to everybody that they didn't have any ideas of their own. Forgers paint copies of Van Gogh and Gainsborough, not of sketches by Watercolour Bert or Aunt Flo from the cottages. Plagiarism is an enduring compliment. People in the field immediately know what has happened anyway, so I regularly have new topics to lecture about and on which to publish papers. Museums have lifted my discoveries and published my results illicitly, authors of papers have borrowed my findings, the BBC has regularly tried to broadcast my research as if it belonged to somebody else (but they never get it right, so they simply provide more material for my presentations).

Using work that isn't your own has now become an industry. If you find it hard to complete an assignment you can buy one online. An assignment can cost £200, which seems a bargain—but they don't necessarily provide good results. A journalist asked for an essay discussing: 'To what extent are women restricted by society in Victorian literature?' The essay that came back began with the words: 'The Victorian era refers to the period when Queen Victoria ruled the British monarchy,' and the text later referred to 'Oliver Twists'. It might be the writer was from SE Asia, because the *Jane Eyre* character Mr Brocklehurst became 'Mr Blocklehurst'. Just a thought.

The best-documented example was in 2019 when a consortium of leading institutes decided to claim they had just made some astonishing breakthrough which I'd famously achieved years earlier. I had discovered some seventeenth-century specimens and reunited them with the microscope used by Leeuwenhoek, the man who had prepared them. Much analysis was done, as you can imagine. This was such an exciting project that the consortium thought they might re-invent it as a new discovery, and claim some publicity of their own. Unfortunately, there had since been over 400 publications on my research, including two highly influential books, and the research is known around the world. This new group took a few blurry pictures as part of a project with a budget of over £700,000. I had taken

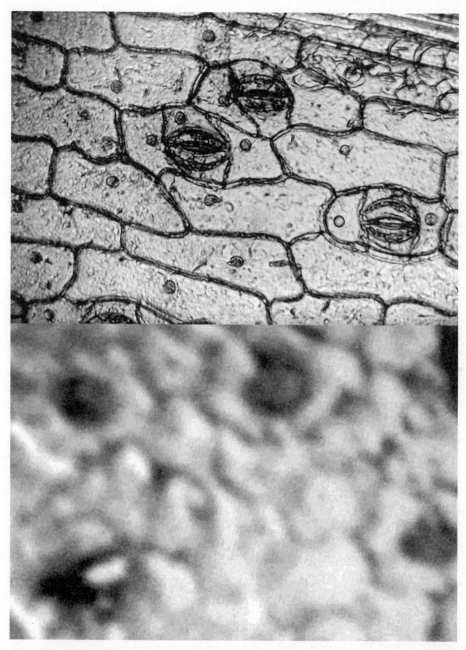

Back in the 1820s, Robert Brown identified the nucleus of the living cell. I restored his original microscope to use and managed to obtain these detailed images with it (*top*). Each rectangular cell has a clear rounded nucleus inside. I mentioned to the BBC that this might make an interesting item for a documentary, so a producer pinched the idea and they made a programme hosted by some dude called Adam Rutherford who tried to repeat my experiment. In this production, you can't see a thing. Brilliant result (*bottom*).

over 5,000 still pictures and more than a couple of hundred videos that were far superior to any of theirs (our budget was £25,000). In addition, we had carried out extensive analysis and specialist identification that was far beyond their abilities. None of this stopped them from suggesting they had achieved something completely new, and sending out enthusiastic press releases to journals around the world. Not a single journal would publish it, because they all knew of the research, though this attempt did become the talk of the town, well, lab. The institutes in the group are the Royal Society, the Max-Planck Institute, the Boerhaave Museum in Holland, and the University of Cambridge, all of whom signed off their claim. They did almost everything right, even having a spokesperson with greying locks and professorial spectacles who spoke confidently (while being completely ignorant of the subject), but made one single slip—the fact that they were copying previously published research was just too obvious. They all know that, if you are too dumb to think up a project of your own, you can steal someone else's. But (and don't forget this) it must be something other people won't detect. Claiming something that's already known around the world is just dumb.

If you want to ensure your publication is easily available online, life has been made simpler through the DOI system. This stands for Digital Object Identifier and it is run by the International DOI Foundation (IDF), which is the registration authority for the ISO standard (ISO 26324). They charge publishers a minimum of $250 per year to join (rising to $50,000 for a big journal publisher handling more than half a billion dollars'–worth of business) and then each published paper is given its own unique DOI—this is a Uniform Resource Locator (URL) that takes people straight to your opus with a single click. A DOI as a URL from the IDF makes life much simpler if you want to check something out, and also if you want to steal somebody else's ideas.

Even when your paper has been submitted, you can still be stymied by The System. Surprising as it seems, even if your paper has been granted a DOI it is not always a guarantee that you've been published. Last year, Amanda E. Spikol of Ulster University submitted a joint paper which was positively reviewed and published in a journal. Eventually they received their customary DOI. They were then told that the journal needed to save money and was going exclusively online. Then they were informed that the journal was closing down completely—and that their paper is being 'unpublished'. That does seem funny, though I had a similar experience. I submitted a paper to the Royal Society and had the usual courteous acknowledgement to say it was going through the editorial procedure … until one of their Experts heard of it. My paper showed that the conventional theories were wrong, and so the Leading Light realised that it had to be stopped in its tracks. This should mean that it was 'declined by the referees' but this would necessitate setting out the reasons for rejection. There weren't any. So instead I had a curious little letter, assuring me that my paper had been 'unsubmitted to *Biology Letters*'. That's unique, as well as ungrammatical. I just had to write a book about it instead, which has been selling well all round the world. What a great world we're in: if a journal goes out of business, they will unpublish your paper. If a referee disapproves of your new ideas, you are thanked for unsubmitting your paper. Remember this. If you disagree with me when you've finished this book, just unread it.

Submissions never have to make sense. In 1996, Alan Sokal, a professor of physics in both New York and London, set out to publish a spoof paper. It was a classical Nonscience experiment, to determine if (in Sokal's words) 'a leading North American journal of cultural studies ... would publish an article liberally salted with nonsense if (a) it sounded good and (b) it flattered the editors' ideological preconceptions.' His paper was entitled: 'Transgressing the Boundaries: Towards a Transformative Hermeneutics of Quantum Gravity' and none of the referees realised it was a joke. Being filled with nonsensical (but complex) paragraphs, it was impossible to tell the difference. John Bohannon published a paper in *Science* showing that a spoof paper could be accepted (in total, by 154 peer-reviewed journals). He invented authors like Ocorrafoo M. L. Cobange by randomly selecting African names harvested from online lists and then adding random middle initials. The Universities he quoted, like the Wassee Institute of Medicine, were invented from Swahili words with African place names. Similarly, a paper entitled 'Fuzzy, Homogeneous Configurations' by an engineer, Alex Smolyanitsky, was accepted by the *Journal of Computational Intelligence and Electronic Systems*. Smolyanitsky did not put himself as the author; that distinction went to two other academics he named Dr Maggie Simpson and Prof. Edna Krabappel ...

THE CONFERENCE HOLIDAY

Conferences are often organised in far-off locations. Christoph Bartneck in New Zealand was invited to attend the International Conference on Atomic and Nuclear Physics in the United States. He says he has virtually no knowledge of the subject, so he claims he put together a presentation using the autocomplete function of the iPhone Operating System. He would begin a sentence with words such as 'nuclear' or 'atom' and then hit autocomplete. The result, he says, was non-sensical: 'The atoms of a better universe will have the right for the same as you are the way we shall have to be a great place for a great time to enjoy the day you are a wonderful person to your great time to take the fun and take a great time and enjoy the great day you will be a wonderful time for your parents and kids,' it read in part, concluding with the memorable words: 'Power is not a great place for a good time.' Within three hours the paper had been reviewed and accepted as a lecture at the conference. All he had to do was register (at the cost of $1,099), and pay for the ticket, transportation, hotel, and subsistence, and he'd be a conference speaker.

People still have to realise that many academic conference organisers are preda-tory travel agents who discovered they could charge far higher prices because their inclusive holiday has an impressive-looking name, and (instead of employing ex-pensive holiday agents) they can use free volunteers to be Conference Organisers who will do all the work for nothing.

Creating text has become much easier since *Nonscience* first appeared. The ideas behind the original buzzword generator and the SIMP text creator have now been digitally automated, and there is a site exploiting these principles. The SCIgen website can create an entire paper for you. For example, they generated a presen-tation entitled 'Rooter: A Methodology for the Typical Unification of Access Points and Redundancy' by the fictitious Jeremy Stribling, Daniel Aguayo, and Maxwell Krohn. Part of it read: 'We can disconfirm that expert systems can be made am-phibious, highly available, and linear-time' and it ran on in this absurd vein from start to finish. You will not be surprised to know that it was accepted as an import-ant contribution by the WMSCI 2005 System conference.

ON THE MAXIMALITY OF CLOSED, TANGENTIAL MONODROMIES

V. WANG, X. WHITE, M. JOHNSON AND Y. LI

ABSTRACT. Let $\mathscr{T}(\Sigma) = M$. In [39], the authors studied free elements. We show that $\nu = -\infty$. Next, this leaves open the question of invariance. In future work, we plan to address questions of solvability as well as invertibility.

1. INTRODUCTION

Recent developments in general logic [39] have raised the question of whether

$$\overline{\mathscr{T}} \supset \frac{\hat{\mathbf{n}}^{-1}\left(\frac{1}{i}\right)}{\mathcal{X}^{-1}\left(\frac{1}{\beta}\right)} \cdots \cdot \pm \mathcal{Y}_{\Xi}(\mathscr{T})$$

$$= \int_{\pi}^{1} \hat{\alpha}\, d\mathbf{x}$$

$$\equiv \left\{ -\mathscr{P}(J) \colon \frac{1}{1} \ni \sum \dot{\phi}\left(\sqrt{2}\right) \right\}.$$

It has long been known that $1i > 1^{-2}$ [39]. On the other hand, in [19], the authors derived unconditionally continuous moduli. In [13], the main result was the com-putation of non-Chern, countable subrings. X. R. Miller [13] improved upon the results of J. Sun by computing equations. It is well known that $C \leq H$.

Those who want a paper that is both impressive-looking and confusingly complex should consult the mathgen site. They provide complete authentic-looking papers like this at the click of a mouse. I was first offered 'Some Convergence Results for Maximal Numbers', then 'Finiteness Methods in Global Number Theory' but settled for 'On the Maximality of Closed, Tangential Monodromes' because of the wicked-looking formulae (or 'formulas', as Americans keep saying).

The most outrageous paper of all? It was called 'Get me off your fucking mailing list' and was written by David Mazières and Eddie Kohler and the entire contents (even the figures) consisted of this phrase repeated over and over again. It was originally intended as a response to people bombarding the authors with messages by email but was accepted for publication in the *International Journal of Advanced Computer Technology*.

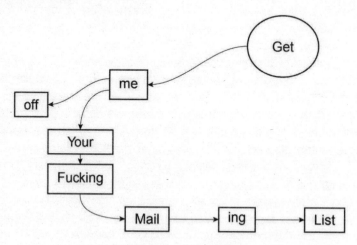

Figure 1: Get me off your fucking mailing list.

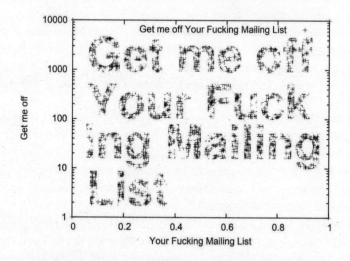

Figure 2: Get me off your fucking mailing list.

The paper by Mazières and Kohler, which found a home in the *International Journal of Advanced Computer Technology*, was entitled 'Get me off your fucking mailing list'. Originally intended as a rebuff to people emailing unsolicited messages, it contained nothing but that same sentence repeated. Like all the best papers, it was illustrated; even the figures retained the theme.

Your paper must have a bibliography, and compiling a list of references has been revolutionised by the web. *Nonscience* recommended you to copy everyone else's bibliography so yours became longer and l-o-n-g-e-r. Now references have a set format (everyone uses the Harvard referencing system) so the citation of Professor Snodgrass would now be:

Snodgrass FJ (1971) Taxation trends—a Prognostication, *J. Internat. Econ.*, **21** (iv) 144–153.

Now, when compiling your references, instead of copying them out painstakingly, all you do is copy and paste everybody else's bibliography and compile them into one huge list of your own. That way, yours is always the longest of all. It's the length of the list they want, not the originality. People copying bibliographies like this is nothing new. In 1887 a paper was written in Volume 26 of *Časopis Lékařů Českých* (the Journal of Czech Physicians) by a microbiologist, Jaroslav Hlava. It had the title 'O Úplavici, Předběžný Sdělení', meaning 'about dysentery, preliminary communication'. He described the management of 60 patients suffering from this disease. Shortly afterwards, a physician named Stephanos Kartulis was compiling his paper on dysentery for the German journal *Zentralblatt für Bakteriologie und Parasitenkunde* (Central Journal of Bacteriology and Parasite Science) and Hlava's paper was one he chose to include. That's where he came unstuck, for he mistook the title 'O Úplavici' for the name of the author, and so 'Dr O. Uplavici' mistakenly entered the bibliography of science as a writer. Others started adding this reference to their own lists, so the non-existent bacteriologist soon became renowned. He was granted a doctorate and was acknowledged by several eminent microbiologists as a source of valuable inspiration. Yet there was no such person, though he was often cited he didn't exist, until the eagle-eyed Clifford Dobell spotted the fact in 1938.

Even more telling is that great reference work *Appleton's Cyclopaedia of American Biography*, published between 1887 and 1889, which recorded over 20,000 names. Scores (perhaps hundreds) are people who never existed. Because the contributors were paid by the number of entries they provided, several of them concocted biographies of non-existent people, and to this day we cannot be sure who is real and who is imaginary.

There are non-existent people in the 1980 edition of the *New Grove Dictionary of Music and Musicians*: neither Dag Henrik Esrum-Hellerup or Guglielmo Baldini are real. In the *New Columbia Encyclopedia*, produced in 1975, there is the life story of Lillian Virginia Mountweazel (1942–1973), a designer of fountains who was born in Bangs, Ohio, and perished in a massive explosion while working for a magazine called *Combustibles*. She had never existed either.

Writers of historical works are regularly ridiculed for their descriptions of non-existent entities like mermaids and unicorns, just as ancient people are ridiculed for

VICENTE Y BENNAZAR, Andres (ve-then'-tay), Spanish geographer, lived in the second half of the 15th century. He published at Antwerp in 1476 four charts, representing the four continents of the world. Unlike Columbus, he did not imagine America to be part of Asia, but represented it as a distinct continent and, what is more remarkable, as a continent divided into two parts by an isthmus. This publication, at so early a date, and before columbus's discovery, has caused much discussion. Some authorities think that Vicente y Bennazar had arrived at the conclusion that America existed as a distinct continent; others, that such an opinion was general among scientific circles in the 15th century; and still others, that he only intended to reproduce the lost Atlantis spoken of by Plato and the ancients.

VOGUE, Jean Pierre de, Flemish adventurer b. in Malines in 1570; d. in Brazil in 1630. He was a captain in the Spanish army when, hearing the fabulous description of the country of the Esmeraldas, he went to Espiritu Santo, in Brazil, and succeeded to the command of the colony that had been founded there. The establishment soon dissolved, as the adventurers were only anxious for riches; and Vogué, having announced that he knew the location of the Mountain of Wealth, was soon at the head of a host, and led an expedition to the interior of the Mamalucos country, wandering for several years through central and south Brazil, and suffering many misfortunes. His confidence in ultimate success was never shaken, but he was abandoned by his followers, and fell at last into the hands of the Charcas Indians. It is supposed that he died during his captivity, although the "Colleccao de alguns manuscriptos curiosos" credits to him the work "Jornada por el descobrimento de las Serras Esmeraldas" (Seville, 1690).

DEATH OF "DR. O. UPLAVICI"

By W. H. Manwaring
Stanford University

IN 1887 Dr. Jaroslav Hlava of Czechoslovakia reported the discovery of amebas in the stools and intestinal ulcers of patients suffering from dysentery, together with his success in transferring the disease to laboratory animals (cats) by intrarectal inoculation with human ameba-containing stools. His paper was published in the leading Czech medical journal of that day, under the title: "O uplavici; Predbezne sdeleni." ("On dysentery; a preliminary communication.")

By some unexplainable editorial oversight, Dr. Hlava's name was omitted in the German reviews of his paper. Credit for this basic medical discovery was, therefore, given to "Uplavici, O" (Dysentery On).[1]

For fifty years international medical science paid homage to the mythical bacteriologist, "Dr. O. Uplavici," ranking him with Pasteur, Koch, and Lister, as one of the outstanding pioneers in modern medical science. It was not until a year ago that the mythical nature of this nineteenth century protozoölogist was recognized by Dr. Clifford Dobell of London, England, and a formal obituary of "Dr. O. Uplavici (1887-1938)" published in a leading English medical journal.[2]

While the creation and perpetuation of the "Uplavici" myth has probably neither hastened nor retarded the development of bacteriological science, Doctor Dobell's obituary may have a salutary effect on future medical historians. There are a number of other minor literary myths still honored in medical research literature, most of them of fairly recent Russian or oriental origin.

P. O. Box 51.

1 Centralbl. r. Bakt., 1: 537, 1887.
2 Dobell, Clifford: Parasitology, 30: 239 (June), 1938.

Invented personalities are scattered through the pages of recognised encyclopaedias. Cartographer Andres Bennazar (*top left*) and Jean Pierre de Vogué (*bottom left*) were both invented by copywriters on piecework who were writing entries for *Appleton's Cyclopaedia of American Biography*. O Uplavici was added to a German bibliography by mistake (this was the title, not the author) and Clifford Dobell wittily published the fictitious scientist's 'obituary' as a way of solving the puzzle (*right*).

Mermaids, classical examples of inventions from the distant past, have their origins in descriptions of basking manatees reported by mariners over 1,000 years ago. Here, Mermaid Melissa, the Manatee Whisperer, drifts towards a dugong to give it a kiss. This gives us an opportunity to review the similarities between the two which gave rise to ancient legends. Melissa is the one on the left.

believing in dragons. Yet fossils of dinosaurs were known to the ancients, and the Komodo dragon is itself a fearsome beast, so a belief in dragons wasn't entirely ill-founded. Mermaids had their origins in the real world too, for the stories were based on mariners seeing manatees breast-feeding their young. It was an exaggeration, but it wasn't entirely baseless.

The unicorn was based on ancient descriptions of the Indian rhinoceros, described by a traveller to an artist who'd never seen one. Portraying the creature like a (familiar) horse was obvious; add a horn on the snout, and there you have it. Muddled and misleading, certainly—but based soundly on fact. In much the same way, modern palaeontologists seriously believe that 100-ton dinosaurs ran about on land and happily copulated without crushing each other to death. Now, that's really ridiculous. We all know that, if an elephant gets into lakeshore mud, it will sink in up to its armpits; yet the palaeontologists' view of dinosaurs perched on a muddy swamp is widely promoted. It does make you smile. In their documentaries on dinosaurs, the BBC include detailed descriptions of mating behaviour, even though nobody has any idea what they were really like. For these seemingly factual documentaries, you just invent anything that comes into your head and people will believe you. Experts can proclaim beliefs that any sane person can see are wrong. Just don't challenge their beliefs, or you will be in trouble.

In contrast to these fact-based accounts, in the zoological literature you can encounter creatures that are completely fictitious. You will read about the mammalian order Rhinogradentia, described in books and articles by the German

If a five-ton elephant tries to walk on mud it will sink in deeply. Everybody knows that. Palaeontologists want you to believe that, when a 100-ton titanosaur walks on a muddy seashore (like this one imaged by Dr Lida Xing from Beijing) it hops across on the surface like a downy duck. This is obvious nonsense, but woe betide anybody who tries to tell them they're wrong.

naturalist Harald Stümpke. Neither exists (the animals and their scholar are rumoured to be the concoction of Gerolf Steiner, professor of zoology at the University of Heidelberg). Did they say that Germans have no sense of humour? Look in their *Pschyrembel Klinisches Wörterbuch Pschyrembe* (the Pschyrembel Clinical Dictionary) and you will find a description of the stone louse *Petrophaga lorioti* which consumes concrete and stone. This fictitious entry was removed in 1996 but users of the book protested, and it was duly reinstated with a supplementary paragraph explaining the involvement of the concrete-eating insect in the collapse of the Berlin Wall.

In my 1971 *Nonscience* book I included some pictures of organisms and living creatures that were published in popular books, but which weren't real. This trend has continued; indeed, it is a lot more evident now than it was then. Search for 'living microbe' on Google images and you will see thousands of cartoons that are nothing remotely like anything that exists. The complexity of real living cells is hardly ever found. If everyday people knew about them they'd only want to know more, and we can't have that.

If you'd like to see a real living cell, then look at a picture taken down the microscope by my colleague Dr Robert Markus and Jafar Mahdavi of the School of Life Sciences at the University of Nottingham. This is a delicious image—full of the complexity of cytoplasm and hinting at the ceaseless activity that goes on inside the cell. The BBC once did a series on cells and created their own version of what a cell is like. Theirs turned out to be a bag of jelly with a blob in the middle. It has nothing to do with real cells; you might as well use a Mickey Mouse cartoon to teach mammalian anatomy.

Remember that broadcast organisations (like the BBC) are our greatest allies. They haven't a clue about what's really going on and the BBC always presents a view to the public that reinforces their ignorance of reality. Those documentaries are mostly on topics like astronomy, space, stress, and dementia, all of which perpetuate the essential superiority of the Expert. Revealing insights people might understand or try at home for themselves (like looking at life through a microscope) are meticulously eliminated from the list. They have pages online that claim to include clips of wildlife: I liked one headed 'animals' which allowed you to scroll on past mammals and lizards, sea anemones and worms ... and there it stopped. Not a single microbe to be seen anywhere. These most important forms of life were simply censored out of the picture. They always are.

Meanwhile, the online academic publishing industry is still expanding with astonishing speed. It used to be costly to launch a journal; but now you can format a cheap online site to look refined and authoritative at the cost of an afternoon's work. Journal sites 'honour' selected individuals by appointing them to their editorial boards, which Experts love to add to their CVs, and which means they carry out costly work for the publisher without being paid. Nobody checks you out. If you don't mind handling editorial duties without pay for a wealthy online

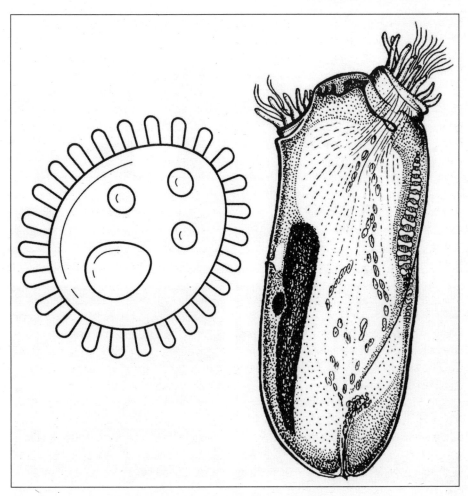

Nonscience insists that the public are never allowed sight of microbes. This is considered altogether too revealing. Who knows? If they knew how exciting it all was, they'd buy a microscope and start looking for themselves, and we cannot have that. Here is a modern drawing of a typical microbe (*left*) which is nothing like anything real, compared with a drawing of *Epidinium* (*right*) which—I hope—is.

publisher, then you can do it too. Becoming a member of the editorial board of a journal is easy. Recently in Poland they invented an author named Anna O. Szust,* a non-existent, poorly qualified academic, who offered to join the editorial boards of various online journals. Within a month 'she' had 40 new appointments.

In 1990, a paper in *Science* showed that about half of all published papers are never cited by anybody and sink without trace. If you are the member of an editorial board you can guide authors, ever desperate to get into print, to support

---

* Interestingly, Oszust is the Polish for 'fraud'.

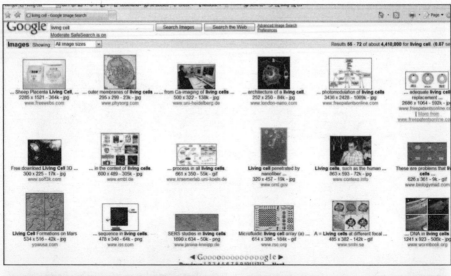

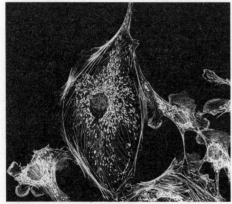

The public rarely catch a glimpse of a real living cell. Search for the term on Google Images and you are met with a vast selection of computer-generated cartoons (*top*). A real living cell is bursting with detail and ceaseless activity, captured in this image by Markus and Mahdavi of the University of Nottingham (*bottom left*). When the BBC launched a series on cells, they concocted this bizarre blob. It is like passing Bart Simpson off as a living human—a brilliant ruse (*bottom right*).

people you like and cite authors you want them to—you can even recommend that they include your papers. But don't take this too far. In February 2020 a bio-physicist named Kuo-Chen Chou was taken off the editorial board of the *Journal of Theoretical Biology*. He had already been removed from the reviewers' panel at *Bioinformatics*. He had been telling authors of papers to cite plenty of his own publications (up to 50 of his previously published papers, in one case). He was found out, only because he stretched it too far. Doing this judiciously happens all the time. You can certainly include your own papers in the bibliography of a paper

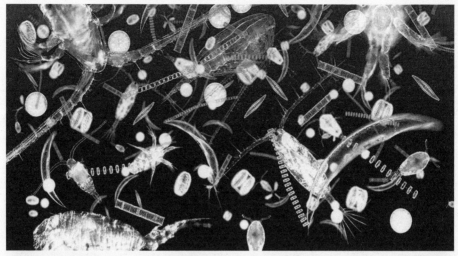

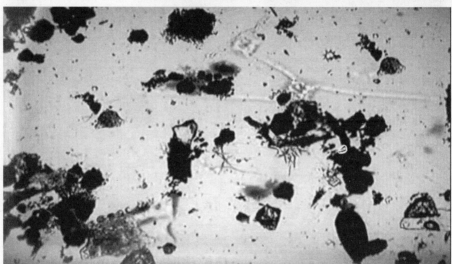

The BBC works hard to keep its public documentaries free of any insight into living cells. It commissioned a film about marine plankton (top) and, in a masterstroke, it took video of views filled mostly with dirt particles (bottom). Since there wasn't much in its samples, it overlaid three separate videos of dirt to make the picture look crowded. Keeping assimilable facts away from the public is crucial, and the BBC excels at this noble art.

you are writing. In 2020 a survey of over 100,000 prominent researchers found that the average paper listed 13% of citations from papers the authors had already published. More than 250 amassed 50% of their citations, or more, by citing themselves. One author had 94% of his own papers listed in his bibliographical references. So do it, by all means, but don't get carried away.

Scientific field	Authors	Percentile, total citations				Percentile, composite index			
		25th	50th	75th	90th	25th	50th	75th	90th
Agriculture, Fisheries, & Forestry	232,801	32	90	255	671	0.997	1.418	1.892	2.394
Built Environment & Design	36,534	17	51	143	370	0.953	1.344	1.821	2.335
Enabling & Strategic Technologies	475,142	23	75	233	678	0.890	1.330	1.807	2.300
Engineering	436,723	18	56	174	499	0.896	1.316	1.794	2.314
Information & Communication Technologies	339,284	20	60	193	574	0.970	1.380	1.862	2.383
Communication & Textual Studies	20,292	12	32	91	240	1.141	1.542	1.995	2.430
Historical Studies	25,277	16	40	105	263	1.138	1.568	2.012	2.429
Philosophy & Theology	13,861	12	32	87	217	1.145	1.558	2.003	2.453
Visual & Performing Arts	3,717	7	17	40	83	0.985	1.316	1.680	1.998
Economics & Business	108,277	28	83	258	708	1.191	1.651	2.194	2.730
Social Sciences	119,260	20	56	158	423	1.159	1.606	2.114	2.615
General Science & Technology	69,789	14	41	122	399	0.735	1.030	1.392	1.760
General Arts, Humanities, & Social Sciences	4,091	11	28	70	158	1.026	1.403	1.810	2.192
Biomedical Research	626,753	68	212	641	1,769	1.095	1.598	2.111	2.660
Clinical Medicine	2,113,734	41	141	467	1,430	0.935	1.420	1.979	2.568
Psychology & Cognitive Sciences	96,159	41	128	403	1,198	1.189	1.641	2.198	2.842
Public Health & Health Services	141,162	31	92	273	785	0.988	1.427	1.949	2.520
Biology	236,108	47	140	426	1,178	1.151	1.603	2.125	2.686
Chemistry	506,526	45	129	362	989	1.057	1.503	1.967	2.467
Earth & Environmental Sciences	223,246	40	126	405	1,192	1.096	1.562	2.120	2.709
Mathematics & Statistics	96,619	18	52	162	457	1.049	1.503	2.059	2.596
Physics & Astronomy	667,255	38	128	480	1,741	1.022	1.495	2.042	2.615
Unassigned*	287,779	2	7	18	42	0.463	0.672	0.985	1.302
TOTAL	6,880,389	29	102	346	1,077	0.946	1.420	1.951	2.513

In order to calculate the c (composite) indicator, any scientist may use the formula $c = \frac{\ln(nc9617+1)}{\ln(nc9617max+1)} + \frac{\ln(h17+1)}{\ln(h17max+1)} + \frac{\ln(hm17+1)}{\ln(hm17max+1)} + \frac{\ln(ncs+1)}{\ln(ncsmax+1)} + \frac{\ln(ncsf+1)}{\ln(ncsfmax+1)} + \frac{\ln(ncsfl+1)}{\ln(ncsflmax+1)}$, where nc9617 is the total number of citations, h17 is the h-index, hm17 is the Schreiber coauthorship-adjusted hm index, ncs is the number of citations to papers as a single author, ncsf is the number of citations to papers as single or first author, and ncsfl is the number of citations to papers as single, first, or last author. The maximum values for these components of the composite indicator are nc9617max = 259,310, h17max = 222, hm17max = 103.9811, ncsmax = 135,334, ncsfmax = 149,125, and ncsflmax = 163,476. For the same percentiles on career-long total citation and composite indicator data split according to 176 subfields, see Table S3.

*Unassigned scientists have no published items that can be assigned to any field. Typically, they have published very few items, and these may be in conference proceedings or journals that are not included in the Science-Metrix classification system.

The data in the Table include all authors who have published at least five items that are classified by Scopus as "Articles," "Reviews," or "Conference Papers."

A study of 100,000 papers (journal.pbio.3000384) showed how often authors cited themselves. This table, they said, shows 'Percentiles of total citations and composite citation metric for each of 22 large scientific fields, career-long data, citations from 1996–2017' and to make it really impressive, they give this stunning formula for how they worked it out:

$$c = \frac{\ln(nc9617+1)}{\ln(nc9617max+1)} + \frac{\ln(h17+1)}{\ln(h17max+1)} + \frac{\ln(hm17+1)}{\ln(hm17max+1)} + \frac{\ln(ncs+1)}{\ln(ncsmax+1)} + \frac{\ln(ncsf+1)}{\ln(ncsfmax+1)} + \frac{\ln(ncsfl+1)}{\ln(ncsflmax+1)}$$

Once publishers have their impressive-looking journals online, and their board of editors, they can then charge each author for the privilege of putting their work on their site. It's a massive confidence trick and, every year, more than half a million papers appear in 10,000 of these predatory journals. They have spectacular titles, like the *American Journal of Theoretical Chemistry and Applied Hermeneutics* or the *International Commentary on Experimental Oncology and Biochemical Cytology* (I just made those up; see how easy it is) and everyone is taken in.

These days people often ask 'How can I be sure I am not being taken in by a predatory journal publisher?' and the answer is 'You can't.' Indeed, many well-known and long-standing journal publishers now charge high fees to their contributors and are parasitic, if not predatory. But don't worry. Nobody's going to check. So long as you have plenty of these published papers, and none of them can be understood, then you're completely in the clear.

You don't have to use online journals to publish, of course. You can blog. Blogging sites will format your random scribbling into something that looks like a professionally produced journal paper. You could even register your own domain (even though *internationaljournal.org* exists, *internationaljournal.com* is up for grabs in 2020). I have known people publish an article on LinkedIn and claim it as a publication, and you can certainly set up pages on Instagram or Facebook if you want to save your money. The golden rule is to add lots of hashtags. At least a dozen of them … anything that will draw attention to your musings.

The final way to make your mark is to publish original thoughts as papers on your personal website. Website design can make anything look authentic and authoritative, and the most trivial of ideas can emerge online as powerful and impressive. When the government initiated lockdown it was on the basis of conclusions from Imperial College—but none was in a published paper. They were just college web pages designed to look official. Another recent example was the theory that the new 5G networks had caused the COVID-19 pandemic. The operating system of 5G ('fifth generation') gives far higher transmission speeds than its predecessors and in 2019 the Trump administration announced they were planning to lead the world with it. Fat chance: as usual, the Chinese are way ahead of everyone else, and their Huawei company leads the world in this new technology. As you might expect, if you draw a map showing where the new 5G has been installed and overlay it with one showing where there has been most COVID-19, then (since most, coincidentally, are found in conurbations, notably in Asia) the two are similar. Bingo! 5G causes the pandemic! This was published on his site by David Icke, who was lambasted round the world for peddling fantasies. Mr. Icke was quick to respond. 'I am not saying,' he announced, 'that 5G is "causing" the virus!' Aha. He is in the clear. But wait: 'I say that "COVID-19" has not been shown to exist! I am saying 5G can trigger symptoms that are being CALLED a virus!' he continued. Hmm. That's problematic. When everyone knows that the epidemic of COVID-19 caused by a virus you can see under the microscope had already killed a quarter of a million people it's hard to say it doesn't exist.

Coincidence offers fertile ground for wild speculation, or 'current research' as we prefer to call it (see pp. 185–188 for more examples). Whenever two phenomena coincide you should simply slot them together and then make an outrageous claim about them (or 'issue the latest scientific findings,' as we like to say). Does wearing socks in bed cause in-growing toenails? Can sitting on a cold wall give you piles? Will a quick shot of bleach rid you of COVID-19? The field is wide open for wild speculation, or 'academic investigation' as it's more widely known. For instance, figures suggest that there has been a dramatic rise in autism over the past decade—just as satellite dishes on the sides of buildings have been appearing everywhere. There's your cause! There's no real connection, but will anybody realise that? Experts will know, and will share a knowing wink, but people like the BBC will lap it up and soon you'll be world-renowned. That's how Nonscience works.

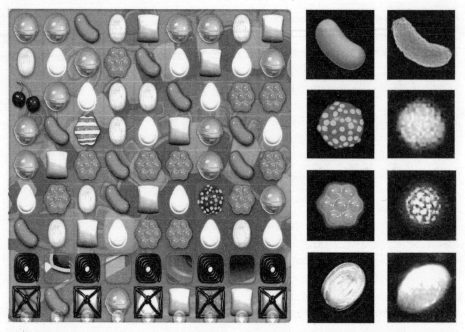

Following the fanciful rumours that G5 installations were causing the COVID-19 pandemic, there are plenty of similar stunts you could pull. The sweets in Candy Crush Saga are obviously a danger to health. They are clearly dangerous germs, with cholera at the top, followed by the papilloma virus, then poliomyelitis, and the oval spores of the much-feared anthrax at the bottom. People who believe these things have one obvious remedy—switch off your phone, stop playing digital games, and start talking to other people. Who knows what benefits might follow?

Think of the other research programmes you could launch. People seemed ready to believe that the radio signals from a digital phone system could trigger a pandemic, so let's find similar projects you could launch. What about Candy Crush Saga? Has nobody realised that the symbolic candies on the screen are actually—disease germs? It's true! The red sausage jelly-bean is the obvious example. It is clearly the causative bacterium of cholera. Candy Crush Saga was released in 2012, when aid workers took the game into Haiti, at a time when cholera was virtually absent. By May and June that year there was a dramatic increase in cholera, particularly among the young. All you have to do is look at the sausage candy and compare it with an electron microscope image of the bacterium and it's obviously the cholera germ. There is also a berry-shaped candy (not Berry the Baker, she's different) which is clearly the human papilloma virus and then there's poliomyelitis and even the dreaded anthrax. You need to turn off your phone. Experts can make the craziest claims, like 'smelling farts makes you live longer'. Don't believe it? Just wait till you reach page 188.

CHAPTER 8

How to Impress Your Colleagues

For ALL the admitted importance of establishing a viable public image, it is equally vital to maintain a propitious working relationship with one's professional fellow travellers. It is partly a question of a well-attuned *qu*FN/*Ere*N cultivation, of course; but there is more to it than that.

PAPER QUALIFICATIONS

As an Attested Expert, you have the right to carry a printed, paper certificate of worthiness. This is what the examinations were for, remember, and it is they and they alone which entitle you to your rank.

You may find, on the first few days of a junior appointment, that some old hands may try to trick you as an Initiate Expert by setting riddles and posers—'go to stores and get two yards of fallopian tube' or 'a bucket of protons', that kind of thing—but do not be perturbed by it. Utilise your reserves of Supracharacteristic Sociotropism and scorn, by implication, such efforts. Always remember a cardinal rule of Nonscience: your certificate is confirmation that you have undergone a course of training spread over a specified minimum period*; it is not an attestation of mere practical ability.

In the normal course of events, bear in mind that the paper itself is your rank, your grade, so to speak; the bachelor degree level (superior by far to technical qualifications such as the Dip Tech or whatever) is the solid baseline of Nonscience and ensures one's admission to the clan. The Master's degree is

---

* Some wags have asked 'what if I passed all the examinations in my first month at college?' This is impossible, since residence in the college course for a predetermined length of time is necessary before the qualifications are awarded. The reasons for this in the context of Nonscience are self-evident.

higher than this, though not as high as the good old PhD. Some colleges rank the PhD as a really rather ordinary degree, as part of their adherence to the Randomisation Principle. The important thing to remember is that as a PhD you are entitled to be called 'doctor', and that goes a long way. Always, but ALWAYS insist upon that. There is then the question of gaining a DSc, which confers no greater social status but does add a touch of exclusiveness within the ranks of Nonscience itself. Sadly, it is hardly worth bothering about unless everyone knows the difference.

Most people don't, unfortunately.

OTHER HELPFUL SYMBOLS

Honorary degrees are conferred on those who manage to gain the highest proficiency in Supracharacteristic Sociotropism. They are awarded according to the personal inclinations of Experts on the staff of universities, and it is well to encourage the correct kind of relationship between yourself and the individual concerned if you wish to get one. It is not enough being efficient, and a mere mastery of the techniques of Expertism alone will not suffice. One must truly display the well-balanced, cultivated development of the SS-mode for success—but do not carry it too far or you could be undone; a certain bending of the image whilst in the presence of those from whom you'd like the degree goes a long way.

Nobel Prizes are excellent tokens of superiority. They often go to those who excel at Infallibalistic Omnipotentiality. It is this development of the IO factor that marks the truly great Expert. Once again, it is necessary for one to be proposed as a candidate to the awarding authority—in this case it is an anonymous group of Swedish Experts—and it is fairly easy to do this by the dropping of hints to influential friends.

Generally speaking it is imagined that Nobel Prizes (which are awarded in fields other than Nonscience, too) are the recognition of peerless ability, and there is no need for you to dispel this helpful illusion. Yet it is the IO factor, not attainment, that matters. It would be invidious for us to examine the merits of actual Nonscience Nobel Prize-winners, no names, no pack-drill as they say; but if we look at literature as a case in point it is easy to see the difference.

Writers who were content to dabble in the hollow practice of what was once termed 'fine' literature have been rightly turned down—examples include Tolstoy, Ibsen and Strindberg. It is those who cultivated the IO factor in their own fields who succeeded, even if their writing was only a secondary interest (and in some cases, it must be admitted, a failing one). Such people included Churchill, Bertrand Russell and Pearl S. Buck. So in Nonscience, remember; adhere to the principles of the discipline; master the intricacies of the IO factor (practise is the only way to succeed) and with a little luck—and a word in the right place—a Nobel Prize could be yours.

FRS

These letters—standing for Fellow of the Royal Society—are sometimes obtained by Experts who master the intricacies of Endofacultatively Reorientational Normoprofessionalism.

It is absolutely vital to understand the dictates of careful self-readjustment that the E*re*N concept implies—and those who confuse the advanced practice of E*re*N with the trivial elementariness of Quasi-notional Fashionistic Normativity (page 35) will not make the grade. A good start here is to cultivate the friendship of someone who is a member of the Athenaeum; from then on, for the E*re*N-positive individual, it is a long—but pretty predictable—haul. It's possible, to add variety, to put RSS (Regiae Societatis Socius) instead. It's bound to add to the fun.

Other Letters after the Name

Joining the appropriate institutes or other bodies associated with your own Expert speciality helps to provide extra letters after the name, which are regarded as being a very good thing by many people (and which if you have enough of them help to give a boost to the Nonscience image).

You will find that several well-established bodies will automatically admit Experts with the right paper qualifications, and they need not attend any meetings (even an initial one for admission), perform any function nor indeed show any sympathy with the aims of the organisation concerned. Holders of even first degrees are as a rule automatically eligible to be a Fellow of the Royal Society of whatever-it-is or a Member (sometimes a Graduate Member, no less) of the Institute of Certain Things, depending on your speciality. Some—such as the Royal Geographical Society—do not require any qualification at all, paper or otherwise; just a few pounds a year can add FRGS to your name. In other cases where it is helpful to boast some honour that cannot yet be achieved, it is always possible to find something similar that—in terms of letters—conveys the same impression. Thus those aspirants who wish to become a Member of a Royal Academy of Sciences may content themselves by joining the Royal Asiatic Society—both bestowing the Letter MRAS—*pro tem.*

For those who have the enterprise to carry this further, there is always AB (Artium Baccalaureus, but it is also a description applied to Able Seamen; and also, for those particularly impecunious, Assistance Board). BDS is the official abbreviation of the Bomb Disposal Squad, as well as Bachelor of Dental Surgery. BL, British Legion, means Bachelor of Laws and Bachelor of Letters; B of H signifies Band of Hope but could always be taken to signify Board of Health. CM, Master of Surgery, actually also stands for Corresponding

Member; Dr, as well as the diminutive for doctors, correctly designates a debtor, too. Aspiring fellows of the Anthropological Society can always make do with the Antiquarians for a few years, and the British Association can supply FBA until you can officially join the British Academy. You may show an aptitude for languages and join the Philological Society until the Philosophical Society takes you in, and still be an FPS meanwhile; and something is always likely to assume it means pharmacy or something like it anyway. MD, Doctor of Medicine, is also an abbreviation meaning 'mentally deficient', which might have application somewhere, and common initials are further shared by the Royal Historical, Humane and Horticultural Societies. The same applies to the Royal Society of Antiquaries, the Royal Scottish Academy and the Royal Society of Arts. It is true, too, of the Royal Society of Medicine, the Royal School of Mines and Regimental Sergeant-Major; and State Certified Midwife shares initials with the Student Christian Movement. There are also odd cases like the Zoological Society, to join which you need (a) either relevant zoological qualifications and experience or (b) neither of these, but enough money to pay greatly increased fees. But the letters are the same either way, and it all helps.

Somewhere along the list of learned societies and organisations are some that will fit your bill. It is hoped that the list of coincident designations, given above, will help you to find the right letters even if the actual body to which you wish to join will not admit you for the time being.

Who will know the difference? Once again (and this is a point emphasised later), it is important to bear in mind paper qualifications and letters after the name as a quite distinct concept from a pilot's licence or craftsman's papers. The former reflect a style of living and certain attitude towards the Nonscience establishment; the latter merely define a certain ability. And bear in mind the emptiness and sad pointlessness of the practical man, the craftsman, the technician, as you reflect on this; they have none of the prestige, the sense of belonging, the purposeful existence at the pinnacles of society that you, as an Expert, enjoy.

Pity them.

FRIENDS IN HIGH PLACES

Even more today than when *Nonscience* first appeared in 1971, you do need the right connections. The best way to network with colleagues is to choose a subject that rates high in the Fashionism charts. When the book first appeared, anything that was an -ology was fine. A colleague of mine made a television commercial in 1987 expressing delight that her grandson (phoning to say he'd failed almost all his exams) had managed to scrape a pass in sociology. 'He gets an -ology, yet he says he's failed!' she wails. At that time, any -ology was the key to success. Not anymore. Those subjects, like geology and biology and geography and archaeology and bryology and zoology and pathology (I could go on) are no longer Fashionistic. First, they are conventional and familiar terms that people know; second, their meaning can be easily understood, and third, the public could imagine doing them for themselves. You should avoid -ologies. A young relative was speaking to her university tutor about courses, and was wisely told: 'Oh— never choose an -ology! You won't get a job with one of those!'

None of the most popular subjects now have -ology at the end: particle physics, genetics, astronomy, theoretical mathematics, nuclear physics, rocketry, computer science, electronics, statistics, systems theory, medicine, philosophy, theoretical physics … that's quite enough. These are mostly subjects that the public cannot understand, and none of them are the kind of study the public could imagine doing for themselves. Note that 'physics' keeps cropping up. This one is a brilliant choice, as is anything with 'theoretical' in the title. Not only does the outsider believe anything you tell them, but so does the grant-awarding authority.

COLLECT THE CASH

For decades there were several Research Councils in Britain, whose job it was to dole out money. It started with the Medical Research Council in 1920 and then came the Agricultural Research Council (1931) and the Science Research Council (1965). Yet there was no sign of a Biology Research Council, just as there is no Nobel Prize for Biology. That was odd: biology is the most important single subject in the world. Through Institute channels and at government meetings I used to campaign for us to have one, but nobody was interested … until 1994. Suddenly a reorganisation was announced that would, at last, make the dream come true. Of course, we were well in the era of Nonscience by this time, where transparency is avoided at all costs, so the new name would not be the Biology Research Council. Too simple. Instead, it would be called the Biotechnology and Biological Sciences Research Council. Brilliant. Means the same but has 19 syllables instead of 8.

Wangling a grant from these Councils is not difficult so long as you follow the rules. Remember that it is their job to hand out cash, so—if they don't make enough awards—they will be in trouble. By taking grants from their expansive coffers you are doing them a favour. You will need the right approach. You should ensure that your head of department, or someone similar, knows people on the grant committee. When submitting the application form, it is imperative that every box is completed properly—even if the question makes no sense, it must still be answered. If it's difficult to know what to say, just bluff using the longest words you can find.

It is the choice of topic that will cause you most problems when applying for a grant. The best thing is to select something you have already completed but which you haven't yet published. Back in the 1960s, after I had visited research centres in the Communist era, I used to lecture in Britain and America on the communist bureaucrats' curious way of doing things. In East Germany they called it *Die Prognose* ('the prognosis')—the applicant had to state what amazing discovery they were going to make, and when they would make it. Can you imagine anything so stupid? Penicillin wouldn't have stood a chance. Nor would jet engines. But that's what bureaucrats want. So, when the Berlin wall came down (we were there that night) a great tide of bureaucracy broke through and swept across Europe, ending up in the Research Councils. Now, everybody has to do it. So keep quiet when you have made a little progress. Apply for a colossal grant, predicting you are going to do what you have already done, and use it (a) to have a great time travelling and (b) to cook up a new idea or two that will be used in the next round of funding. This system cannot fail.

Applying for grants takes time. I was once awarded a small grant by the Wellcome Trust in London. With it came forms that had to be completed as the research went along. I worked out one day that the amount of the grant just about paid for the time consumed in filling out the forms. This is a general rule. Some surveys have shown that researchers after grants can spend 40% of their work time in seeking funds. If you're wise, you can reduce this enormously—mainly by applying for lots and lots of money in a single award. It is a curious fact that huge grants are more easily obtained than small ones. Meanwhile, the time taken in deciding which grants to provide can itself consume enormous amounts of money. The Natural Sciences and Engineering Research Council in Canada has spent more than C$40 million in one year administering its basic 'discovery' grants. If they had simply divided up that money among all the applicants, they could have had C$30,000 each. It is worth looking for quirky grants; for instance, in Glasgow there is a fund that awards grants—but only to students with the surname Graham. I don't know what happens if it's hyphenated.

Committees don't understand research and will give grants worth millions for topics they don't comprehend. These days it is smart to boast of ignorance of such things, and a lack of basic understanding in this area is considered perfectly acceptable in modern society. People will smilingly tell you they 'can't understand maths', or that they 'don't know anything to do with physics', or how 'research is completely beyond me'. This is what we need to encourage. They would never admit 'I don't know anything about Shakespeare' or 'I'm completely ignorant of music' and would never say 'impressionist painters don't interest me' but for the topics that nurture Nonscience they happily confess their lack of knowledge. They just love to know that wise people control the world with subjects that the public cannot grasp. Remember that when choosing your career. So long as people are confused about what you do, your success is assured.

CHAPTER 9

Fame at Last

So here you are: a lifetime of close training and discipline, a secure job in one of the best of the modern professions—Nonscience. You have a dozen or so papers behind you and perhaps a book too. And all this time you have managed to steer clear of the pitfalls that dog your path—being too familiar with Lay People, trying to explain reasons, rather than sticking to hard-and-fast facts, that kind of thing.

It is not long before the wider public hear of you. This may well be when an organ of the mass media contacts you direct to ask for a contribution. The main categories are:

- NEWSPAPERS. They often publish 'special supplements', devoted to a single topic—nuclear energy, laboratory apparatus, the drug industry, etc.—and intended to encourage advertisers' support. Most companies will happily purchase advertisements in such supplements for reasons of prestige and though it may occur to you that the benefits of doing so must be marginal at best, it would be churlish of you to say so at the time. If you are asked to write an article for such a supplement, just remember that the fees are moderately good, the articles are needed to fill in the space between the advertisements and if you don't write one somebody else assuredly will. You were approached, remember, and that means that someone has noticed your work—fame is already within your grasp.
- WEEKLY MAGAZINES. These include the whole range of general-interest publications, some scientific, some superficial; but the kind of article they want will be indicated by the editor—and whatever it is, do make sure that it is sufficiently eye-catching and sensational to catch the public interest. One should fully utilise words such as 'astounding' and 'sensational', and should modestly suggest that your own approach to the subject is clearly the only one that could be taken seriously, out of all the available choices. Make sure that the article reveals you for what you are—a successful Expert, a

person to be respected, trusted even—and eventually the television people may become interested. But more of that later …

- POPULAR ENCYCLOPAEDIAS, etc. These were, traditionally, published as well-bound glossy volumes but in the 1960s there has been a tendency towards weekly serial publications. They are known in the trade as partworks. There have been many of them—covering subjects ranging from aviation to sex, from witchcraft to motoring—and they often approach Experts for their contributions. An interesting article can be in front of thousands of families, some of whom will be TV directors and suchlike (who often buy this kind of thing), so take care that it is interesting, provocative and full of hard facts. Who knows what new avenues can open up as a result?

In writing for media such as these, it is vitally important to get your priorities right. First, do not forget the purpose of Nonscience (its own promulgation) is the primary aim. Second, do not attempt merely to 'explain' subjects in an amorphous, intellectual manner. The topic should be outlined with data, facts and figures; if the background is all accounted for and a full explanation given

THORIUM CHAINS – Dr. Albert Crewe of the University of Chicago displays a photograph showing for the first time individual atoms within a molecule. Enlarged five million times are Thorium chains in which the small white dots represent single Thorium atoms. The picture was made with an electron microscope that Dr. Crewe designed and built He expects ths tool to be useful in research on a cure for cancer.

Note how efficient government-sponsored publicity transforms an abstruse electron microscope into something that affects the Ordinary Lay Person (page 245).

then there will be no point in anyone asking you to comment further—you have destroyed your own market! Keep the trump cards up your sleeve, and never, ever give away every secret. That would only degrade the discipline as a whole, and you into the bargain.

Sometimes for some reason or another (often due to a failure to observe the basic principles of the game) no-one, still, takes any notice of your existence; the media remain silent, the newspapers don't refer to your Expert knowledge, the whole press and broadcasting machine continues to pass you by. It is at this stage that you can with some advantage organise a special press day in the form of a visit to your establishment. The most directly effective method is as follows.

Bearing in mind that there is absolutely no point at all in sowing seeds on infertile, unprepared ground, begin by sending a series of press releases to the national media, and also to the local newspapers and broadcasting studios if that seems appropriate in the context. A press release should conform to several basic rules:

(a) It should say as little as possible about the actual work.
(b) What it does say it must say with boundless enthusiasm.
(c) It must embody certain key references to Fashionistic implications of the work. They are listed on page 169.
(d) It should at all times seek to make the research programme relevant to the most liberal and optimistic interpretation. It is hard to put this precisely into words, but there are many examples around:
  i.   The University of Chicago, in 1970, found that one of their researchers had made a novel type of microscope with all manner of interesting new facilities, including mass spectroscopic analysis and a high-resolution scanning facility. And just how interesting is that to the public? Not in the least! But a facility of this sort could be used for the examination of organic molecules, couldn't it? And living cells contain chemicals of an organic nature, don't they? Furthermore, cells *can become cancerous* ... and there you are: THE SENSATIONAL NEW MICROSCOPE MAY BE OF USE IN THE FINAL BREAKTHROUGH AGAINST CANCER.
  ii. Another specialist, by using well-known tried-and-tested ideas in microsurgery on free-living protozoans, succeeded in transferring cell membranes from one cell to another, and implanting nuclei too—rather like rose grafting, and in principle the same kind of manoeuvre that has been carried out in countless earlier experiments. That wouldn't attract anyone, now would it? But these are *living cells* (an important key phrase) and the end-product was certainly different from the raw material, even though none of the cellular components had been changed

or altered in any way. And there is the clue! THE FIRST SCIENTIST TO CREATE A LIVING CELL (abbreviated later to CREATE LIFE IN A TEST TUBE) claimed the press. Full marks there too.

It is likely that reporters, given these press releases by the news department editors, will ring you up to enquire more about the development. This is where the correct use of Nonscience-consciousness is necessary. Your secretary/technician should always ask such a caller to hold on for a moment, and should then

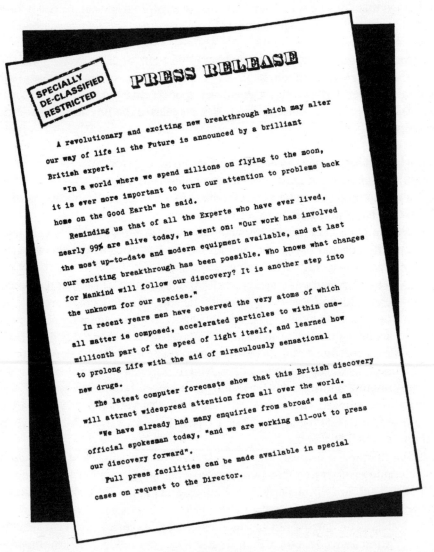

A universal press release, suitable for all fields of discovery and invention. Simply insert your own name and the laboratory's address at the bottom before duplicating.

come back on the line to say 'I'm sorry, but Dr Virus* is very busy in some crucial experiment right now. Could you perhaps write him a letter?' This is a very necessary ploy because newspapers and broadcasting organisations invariably like to have a topical theme in any item. It is known professionally as a 'peg'. The reporter, faced with the actuality of an experiment in progress, will either ask to come round or will plead for just a brief word with you. It is important to keep them waiting on the phone for perhaps five minutes or so, during which time their eagerness will increase a hundredfold, before speaking to them yourself.

SOME KEY PHRASES

Revolutionary breakthrough

Living human cell

Basis of Life Itself

Secret of the Universe

Immediate autopsy (note: the correct term is 'necropsy', dissection of the dead—'autopsy' means 'self-dissection'—but is it important to adhere to Nonscience orthodoxy)

Hitherto unsuspected

Vitally alter our way of life

Completely unknown

Masterminded by an Expert

Astounding discovery

Human sexual urges

Miniscule (correctly spelt 'minuscule', but Nonscience convention is in line with miniskirt, minicar, etc.)

Creating life in a test tube

Impressive new development

Artificial life

The computer

A million times greater/smaller/ bigger/faster

Nuclear energy

Sexual abnormalities

Women's liberation

Pollution/ecology

Knickers

The socialist ethic

Sub-atomic particle

The vital terms:

  Vomit,

  Fucking,

  Pus,

  Spew, etc.

Step into the unknown

Ninety-nine percent

Official spokesperson

Incalculable benefits

Starving millions (feeding/housing of, etc.)

Top secret

Latest

New

---

* Your own name would ordinarily go here instead, or did you work that out for yourself? Well done, Experts!

The concept of a 'peg' on which to hang the item is an important one to bear in mind. It is close to the Expert's concept of quFN orientation in some respects. There is, of course, no particular merit whatever in topicality for its own sake, but to the mass media 'pegs' are highly Fashionistic. To illustrate the point:

Case A. Dr Virus has discovered a new chemical material that in rats can prevent the development of cold symptoms after an infection with the correct germ. It is late spring and a sunny afternoon. HE HAS NO CHANCE WHATEVER OF PUBLICITY FOR THIS ONE

Case B. Dr Virus sends a press release to the media to coincide with an official (ministry) proclamation about influenza; it is just after Christmas when the 'flu season is at its height and nothing newsworthy ever happens. The press release says nothing new, but points out Dr Virus's views on colds and influenza. THE PEG OF THE MINISTRY STATEMENT AND THE SEASONAL TOPICALITY ENSURE DR VIRUS A PLACE ON TELEVISION PROGRAMMES LIKE *24 HOURS, LINE UP, NEWS AT TEN*, AND *BLUE PETER* WITH LUCK.

In short, good item but no fashionistic 'peg' = no publicity.
BUT run-of-the-mill item, topical 'peg' = high publicity value.

One can always manufacture a 'peg', of course; and Dr Virus would have been assured of success if he had sent out a press release which began with the words:

AS MEDICAL AUTHORITIES PREPARE THEMSELVES FOR NEXT SEASON'S INFLUENZA OUTBREAK, WHICH MAY WELL CAUSE MORE DEATHS THAN ANY IN HISTORY, A FAMOUS ENGLISH EXPERT IS ABOUT TO ANNOUNCE A REVOLUTIONARY NEW DRUG THAT CAN AT LAST BRING US THE BREAKTHROUGH AGAINST HUMANKIND'S GREATEST ENEMY—A SCOURGE THAT MIGHT OTHERWISE MEAN AN END TO EVERY MAN, WOMAN AND CHILD ALIVE TODAY.

With devotion and adherence to these principles in detail, it is possible to find the entire might of the mass media turned in your favour. The swift rise to fame of Dr Christiaan Barnard is a case in point. It demonstrates how the unreasoned hysteria of the media of communication can easily get out of hand.

His decision to carry out a heart transplant operation on the late Mr Washkansky came at an excellent time. Major transplant surgery had been carried out for years, but it had not been Fashionistic then. The first experimental heart transplant, in fact, dates from 1905! A lung transplant in an American convict took place in 1963 without any significant interest being shown, even though it calls for more drastic surgery. Animal organs had been transplanted into humans in operations too.

The effect of all this had been to create an undercurrent of interest in this subject, which by the end of the decade had grown until it was highly Fashionistic.

It was then that Chris decided upon his epic operation. Was it because he was the world's most experienced transplant surgeon? Far from it: other teams elsewhere had carried out far more experiments on animals, and with greater success too. Was it because he had a tried-and-tested technique, better than the others? We must admit that this was not so either. The technique used occurred to him in bed on the afternoon of the operation and had never apparently been tried (by Chris or any of the others) before. Was it because his technical facilities were the world's best? Indeed not: the donor heart–lung machine was, in his own words, 'the old one of many memories I had brought back from Minneapolis.' Neither was it because of surgical skill of a world-beating

Dr William J. Kolff and his artificial transplant heart, dating from 1961—long before Press Fashionism had developed for this particular topic. Tough luck there, Bill.

standard. Chris was already suffering from arthritis, which was in itself dictating the end of his surgical career, and the operation was not entirely free of what he later called 'stupid' mistakes that nearly cost the life of the patient.

Getting the patient to agree to such experimental surgery is itself something of a task. But Mr Washkansky was, we must remember, a member of the Lay Public who naturally embodies a certain reverence and respect for Medical Experts as a whole. He was, as is any man with chronic heart failure, willing to take even a remote chance. And Chris had told him that he stood 'an 80% chance'.* So it was not difficult to ensure his cooperation.

Eventually the operation was completed, and the news was broken to the world's press. With a high-Fashionism climate of opinion (see graph on next page) it obviously was a favourite subject for reportage—but that alone isn't quite enough. It needs the added ability to be amenable to press interest, to be friendly and polite to reporters, generally to 'melt' a little for the Media—in short, to utilise the E*re*N principle, as we in Nonscience prefer to call it.

It was Christiaan's ability to do just this that capped the whole performance. As the subject cropped up in the media, with added enthusiasm resulting from the factors already elucidated, a snowballing, mushroom growth took over. He was photographed with a range of assorted people, from the Pope to Sophia Loren; an LP was prepared and there was talk of a film on THE BARNARD STORY ... meanwhile other surgeons elsewhere, on the crest of Fashionism, suddenly began to do similar operations and at the same time hordes of Experts the world over began to collect substantial fees for commenting, writing or broadcasting about the Significance of the Operation itself.

But of course there is no answer to the rejection problem and infections soon took over in Washkansky's body. Had they been latent, hidden in his body? Or were they introduced by reporters who were admitted to the ward area? We shall never know ... All that is certain is that the Media were given a generous insight into the experiment. Mrs Washkansky's first visit to her husband after the transplant was—though she did not know it at the time—recorded in detail by that ubiquitous microphone. And Chris was soon off around the world, doing radio and TV shows, visiting nightspots and completing an exhausting round of appearances.

At the end of it all the subject of transplantation had been boosted to unprecedented heights of Fashionism. Many operations—heart valve replacements, etc.—that had been routine for *years* were suddenly written up in the press under banner headlines. It all faded away within a few months; in fact, by late 1970 transplantation surgery had almost lost Fashionism altogether. But while it lasted it was *most* impressive.

---

* The patient took this, according to Chris, to mean 'an 80% chance of success'. In fact, it meant 'an 80% chance of getting out of the operating theatre still alive'. But *why split hairs*?

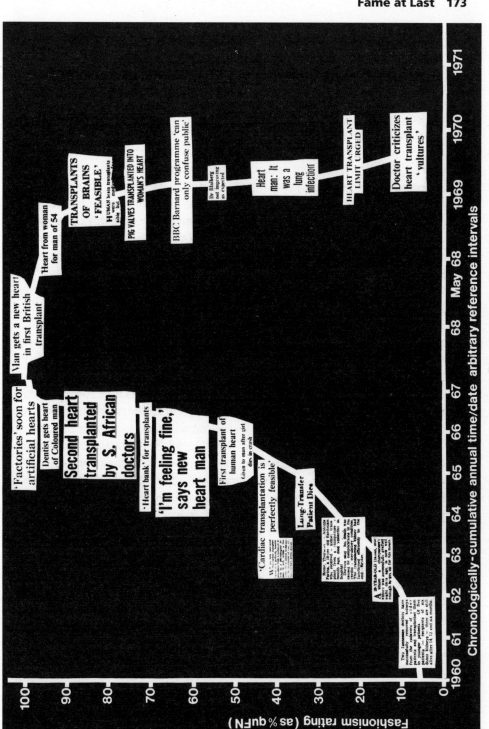

How the press produced the rise and fall of Transplant Fashionism.

Well, Experts, that is what can be done if you really hit the big time. However, we must be realistic—most research projects will never reach those proportions. Yet with careful handling of the press a gratifying (if modest) amount of publicity can be yours.

Assuming that you have, as earlier paragraphs instructed, carried out a successful groundwork scheme (so that in the minds of most editors you are now Fashionistic) you are ready to have a press day:

- INVITATIONS will be sent to newspapers, periodicals, radio and TV. They will be accompanied by a press release extolling the virtues of your organisation and indeed yourself. It is important to aim the appeal of the event mainly at the newspaper people. This is because TV and radio producers get their topical (i.e., Fashionistic) ideas from the press—broadcasting networks recruit from the newspapers, and do not train reporters from scratch themselves—and in this way newspapers really set the pace for what is to be covered, and what isn't. Producers throughout the world open their papers, scan the columns and then announce 'Right, we'd better cover this event then, people' and it is as well to bear that in mind when you are planning the orientation of your invitation list.
- FACILITIES should include cleaned, sparkling laboratories with all available apparatus ticking, humming and blinking away like mad. It is not necessary for it to be *doing* anything. The walls should be covered with neatly Letrasetted charts, diagrams and obscurantific explanations that only Experts can unravel (and then, remember, not *too* much at a time). An impressive office should be labelled FOR PRESS USE ONLY and if a couple of telephone cubicles are available—labelled in the same way—the journalists will be clamouring to use them and send out stories first. The element of competition ensures that each will want to be more topical, more newsworthy, more Fashionistic, than the rest. With luck, none of them will have time to stop and wonder whether they ought to be bothering at all in the first place.
- REFRESHMENTS must be liberal. Draught beer, scotch and gin are the drinks that will be requested. Two or three barrels of beer and two dozen bottles each of the spirits will do. Ginger ale is vital for the whisky, tonic for the gin; ice with both is popular. So get those specimens *out* of the freezer and make some cubes in advance. Do not forget the lemon, either. These are all purchased on grant funds as a necessary item of expenditure, of course.
- FOOD should be in the form of little sandwiches topped with parsley or cress. Canapés are popular and so are slivers of smoked salmon and portions of imitation caviar (lumpfish roe) as no-one will tell the difference and you can always charge up for the real thing—some 800% dearer—on the petty cash.

INSTITUTE OF TECHNOLOGY,

TOWNHAM ~~TRAINING COLLEGE~~
43a, HIGH ROAD,
TOWNHAM.

Dear Albert,

Harry and I would be glad if you would like to drop by one afternoon this week. Our experiment is working At Last.

Should be worth a little Mention for your Man About Town column, we thought!

Love to Mabel and all,

Pete.

Attention: assistant News editor

How not to inform the Press.

RESTRICTED: To the Senior News Editor

Hope you can make it, Al!

Invitation

PERSONAL AND CONFIDENTIAL

The Senior Research Director ....

requests the pleasure of your company at an Exclusive Demonstration of an important New Breakthrough on

13th August at 1350–1735 hrs

Full Press, Radio and Television Facilities will be provided, and telephone lines for the exclusive use of the Press have been specially installed.

Drinks and Buffet service provided from 12.30 p.m.

A Car Hire service has been made available.

For security reasons this Invitation Card must be carried and Identification Badges with Radiation Detector will be issued at the Lodge.

SECURITY NUMBER: AB 924/339A

The correct way of doing these things.

- DEMONSTRATIONS are so variable that it is impossible to give general examples. However, they should be:
  - i. *gory*, to attract attention;
  - ii. *photogenic* since photographs always attract attention in the press; and
  - iii. *unique* or at least described as such in order to assure Fashionistic Newsworthiness.
- REMEMBER to treat all guests as Special Visitors. This is the E*re*N-factor coming into its own again (page 108) and it will guarantee results. From then on it's up to you.

Finally, a word on what happens to the information after the media have it. Newspapers and magazines simply write articles about it, and the success of your campaign is calculated solely on the basis of the number of square inches consumed in the publication. It is space which matters, and not accuracy. There is an unwritten tradition that whatever you have said to the press will always appear in a dramatically different light when the article has been written.

Newspapers, for instance, always get it all wrong. There is a tradition about this which it is unwise to question and quite impossible to alter. Developments are generally written up as (a) death rays, (b) super-germs, (c) keys to life or the universe, etc. (page 169 for an exhaustive list). Each item is a 'breakthrough' towards the 'ultimate'. Do not be perturbed by it. That's how life is. Be content to be flattered by the inches that your story covers, even if it does say that you have learnt how to cure a disease that actually you have only just invented.

Alternatively you may be interviewed for radio, in which case a little judicious editing of the tape can be done to extract the juicy morsels for public consumption (thus in answer to a question such as: 'Tell me, Professor, is there a risk to human life from your discovery?' you might give the answer: 'I don't know whether you could put it as strongly as that; as far as we know it's entirely harmless' but the edited version could well be: 'I don't know …'). But do not be upset by all this. If the item isn't made appetising in this way it won't be broadcastable at all, and then where would be your quest for fame and power, eh?

Television interviews operate in the same way, unless you have to go to the studio. Beware of the impedimenta of a studio. Only the slim person with a hearing aid (known as the *floor manager*) is allowed to touch any of it. Only the novice sits down, moves the microphone into a 'comfortable' position only to find that he ends up mouthing his discovery silently to the viewing millions when the moment of truth dawns. You will be asked for a little 'level', i.e., a few sentences for the sound mixer to adjust the setting on your microphone. Most people, when asked to say a few words, invariably whisper: 'w-w-w-what shall I say—I mean, anything? Like w-w-what?' to the accompaniment of titters from everybody else. The hardened Expert who wishes to seem a pro, on the screen simply talks away, using SIMP sentences (page 120) or something else that no-one can understand, thereby showing that he is (a) *au fait* with what's going on and (b) immensely clever. Don't peer at yourself on the monitor either.

Your name is virtually in lights already; stand by for Fame!

— ₥ —

COLLEGE INDUSTRY

The picture of Albert Crewe and his microscopic view of atoms was a textbook example of how to exaggerate your research. It didn't end there. More than 40 years later the same picture was still being used, this time as a clue to clean energy. Imaging thorium atoms has nothing to do with either field of research; it simply shows how publicity matters more than reality.

With all their new money from the young, universities have been a growth industry. The original book mentioned how their numbers were expanding. In the

Is thorium the answer to the world's energy woes?

An Irish duo have set up an ambitious kickstarter campaign for a documentary on the issue, they told TheJournal.ie.

May 26th 2013, 8:05 AM 👁 11,975 Views 💬 30 Comments

f Share 105   Tweet 42   ✉ Email 45

TWO IRISH FILMMAKERS want to know if thorium could be the answer to the world's energy woes – and they have launched a Kickstarter campaign to raise enough money to finish a documentary on the topic.

Frankie Fenton and Des Kelleher want to raise £40,000 so that they can finish post-production on The Good Reactor, their documentary looking into the alternative nuclear fuel and whether it can generate clean, safe energy.

University of Chicago Physics Professor Albert V. Crewe points to a photograph which enables a single atom to be seen within a molecular structure for the first time. Enlarged 5 million times are thorium chains

This picture showing thorium atoms featured in the original *Nonscience* (page 166) as offering a cure for cancer. It didn't offer anything like that. It appeared again in 2013, this time as a source of clean energy. The photograph hasn't anything to to do with that either. What next? My money is on it treating dementia. Or perhaps thorium can eat plastic.

10 years before *Nonscience* was published, the number of universities doubled from 22 to 44. The advice in my book was slavishly followed, so now there are 132. Some were created from existing colleges and training schools; others (like the polytechnics) were already of university standard and simply changed their names. Many were new builds with lots of glass. Others, like the University of Wales, were composed of several colleges so each separate college was now called a university instead. The boss, who was appointed as a Principal, started calling themselves 'Vice-Chancellor' instead and there you are—job done.

The name change idea is at its funniest in London. London University embraced 18 separate institutions (currently they have over 160,000 students on their list) and they became addicted to name changing. They were originally called London University from 1826 to 1836; then they transmuted into University College, London (1836–1907), from there to University of London, University College (1907– 1976), and eventually to University College London since 1977. As you can see, there should be a comma in there somewhere but at the turn of this century that kind of punctuation lapse had them alter the names of all their constituent colleges to make them sound grand. Since those other colleges had changed from 'college' to 'university', London decided to follow the advice in *Nonscience* by simply dropping the word 'college' altogether. The result is an ungrammatical mishmash. For instance, Royal Holloway College was one of their constituent institutions. Now it is listed as Royal Holloway University of London. Similarly, we have the bizarrely named Queen Mary University of London. Neither names make any sense. St George's, University of London, at least has the comma; and (explain this if you can) they retained University College London and King's College London. To an outsider, this seems uneducated and confusing—and therein lies the truth. Note what I said above: 'explain this if you can.' You can't. That's the whole point. Nonscience exists, not to explain, but to confuse and disenfranchise those who are clearly too stupid to understand these things. Experts never explain! They just *say* things.

Universities have their own way of acting which nobody is ever allowed to challenge. Experts acknowledge that fact. Like other traditional seats of learning, Christ Church, University of Oxford, has an arcane form of management that includes pay structures which are completely out of touch with present-day realities. The head of the college is the Dean, Martyn Percy. He had only been in post for four years when he thought he could sort it out. He also wanted to open admissions to youngsters from state schools. Naturally, the other dons decided that they wanted things to stay the way they were. There is a way of dealing with these things. A standard letter would be: 'It has come to the attention of the Fellows that our Dean is proposing a series of fundamental organisational and operational modifications to our traditional modus operandi. We wish to state, with great respect, that, after exhaustive consideration in great detail, it is clear that such alterations would detract from the traditions that we revere and which our students hold dear, and therefore we are resolved that these proposals should not be introduced within the foreseeable future.' That is precisely how the public imagine such things to be handled.

However, that is only for appearances. Fellows aren't really like that. Internal messages said that the Dean was a 'manipulative little turd' who was 'thick', 'nasty', and had a 'personality disorder' making him act like a little Hitler. 'Please, please, get rid of him,' implored one of the dons and another said: 'Yes, he's got to go,' while describing him as 'nasty and stupid.' One note asked: 'Does anyone know any good poisoners?' and another added: 'Think of the Inspector Morse episode we could make when his wrinkly body is found!' The College joined in, reportedly spending £2,000,000 of their charitable funds on an internal action against Dr Percy.

Christ Church is a charity, and wealthy enough for such things. The college was founded in 1546 by King Henry VIII and currently has assets of £550 million so they can certainly afford it all, but annoyingly, the College has been reported to the Charity Commission because it is thought that the expenditure is not a justifiable use of charitable funds. Dr Percy fought back and is said to have already spent £400,000 on defending his stance. Supporters of the Dean have raised £100,000. Obviously, none of this is supposed to be known outside the College. It is nosey reporters, who should just mind their own business, who have brought this out into the open. Now everyone is embarrassed. You should never think of modernising an ancient university. Arcane is what they're meant to be. That's all part of the mystery. And can you all remember in future that internal disputes are meant to be just that, internal? This will blow over, certainly, but it remains of paramount importance that the public are kept at bay. Universities are not afraid of change, but the changes they need are those that promote the world of the Expert, not unwanted intrusions that might let the cat out of the bag.

It is the commercialisation of universities which means that they can pay their senior staff astonishing amounts of money. If we look at the top 24 universities in Britain (the so-called Russell Group*) currently 508 of their staff earn more than £160,000, which is what the Prime Minister earns. The previous year it was 428 people, so you can see that the numbers are steadily increasing. You can make more in a university than in business. Oh, hang on, a university is a business. I was forgetting.

These changes all hinge on promotion and PR, and the general-purpose press release published in the original book could still be used today, in fact, something like it regularly is. And the advice about making an announcement official by organising a formal press conference—with lots of alcohol and comestibles!—is

---

* The Russell Group was set up in 1994 by 24 universities who agreed between themselves that they were the best. Why the name? Was it to commemorate somebody like Lord Russell, doyen of philosophers? It was too early to be named after Russell Brand, though it's an interesting thought. No, the name comes from the Russell Hotel in London where they held their inaugural meeting. Note that they met where all those attending could enjoy lavish hospitality. Start as you mean to go on.

now a fine art. There are organisations that will stage a huge party to make your announcement unforgettable. Critics describe this as a 'cheap scam' which is wrong. It is *not* cheap.

HOLE LOT OF MONEY

One great recent example was the announcement that astronomers claimed to have obtained an image of a black hole in 2019. Everyone has heard of black holes, though the public, naturally, cannot understand what they are.*

Public interest in black holes had given rise to numerous published impressions of exactly how one would appear. These stunning examples appeared in magazines around the world and were widely screened on television. They were all over the Internet. The public, already confused by what a black hole might be, were left impressed by the ingenuity with which astronomers were pursuing their amazing investigations.

The image was a fuzzy picture showing little of interest, and nothing that was new. In the old days you'd show this to your colleagues and they'd chuckle at how bad it was. But not now! A huge press conference was arranged, using an invitation remarkably similar to the example in the original book. People turned up on the day, from all around the world, though half the seats were empty. The stage was refined and elegant, with a panel of neatly coiffed Experts sitting languidly on expensive chairs. The event began with statements from the European Commission podium. Said one speaker: 'To take this picture … you need people from 40 different countries, you need people from all over the world and we have put together over €44,000,000 (44 million euros).' Then he announced the imminence of the Big Moment which, he assured us, was being shared with so many people around the world, and so many media people at seven simultaneous press conferences. Then he introduced the man who was due to unveil the new discovery, which he said was in a galaxy called M87 that lies 500,000,000,000,000,000,000,000 (five hundred billion billion) light-years, or kilometres, he corrected himself, away from Earth. Then the indistinct image flashed up on the screen. The room was dark, and the media audience sat in stupified silence for several seconds at the stunning

---

* Black holes are obvious to a child. Matter is attracted to more matter by gravity (which is why we stay stuck on the world). As matter gets closer together, the strength of the gravitational attraction increases. Well, obviously, if this went on until bits of actual atoms were crammed together, the gravity would be so strong they would all crowd into a smaller and smaller space as the gravity became stronger and stronger … until the gravity was so strong that it would draw in everything, and even light couldn't escape. There you go: a massive collection of super-heavy matter which (because light never escapes) you cannot see. They were first named black holes in 1953, from the Black Hole of Calcutta, an infamous Indian dungeon used by the British, from which prisoners never re-emerged.

The European Commission, European Research Council, and the Event Horizon Telescope (EHT) project will hold a press conference to present a groundbreaking result from the EHT.

- **When:** On 10 April 2019 at 15:00 CEST
- **Where:** The press conference will be held at the Berlaymont Building, Rue de la Loi (Wetstraat) 200, B-1049 Brussels, Belgium. The event will be introduced by European Commissioner for Research, Science and Innovation, Carlos Moedas, and will feature presentations by the researchers behind this result.
- **What:** A press conference to present a groundbreaking result from the EHT.
- **Who:** The European Commissioner for Research, Science and Innovation, Carlos Moedas, will deliver remarks. Anton Zensus, Chair of the EHT Collaboration Board will also make remarks and introduce a panel of EHT researchers who will explain the result and answer questions: Heino Falcke, Radboud University, Nijmegen, The Netherlands (Chair of the EHT Science Council). Monika Mościbrodzka, Radboud University, Nijmegen, The Netherlands (EHT Working Group Coordinator). Luciano Rezzolla, Goethe Universität, Frankfurt, Germany (EHT Board Member). Eduardo Ros, Max-Planck-Institut für Radioastronomie, Bonn, Germany, (EHT Board Secretary).

RSVP: This invitation is addressed to media representatives. To participate in the conference, members of the media must register by completing an <u>online form</u> before April 7 23:59 CEST. Please indicate whether you wish to attend in person or if you will participate online only. On-site journalists will have a question-and-answer session with panellists during the conference. In-person individual interviews immediately after the conference will also be possible.

The conference will be streamed online on the <u>ESO website</u>, <u>by the ERC</u>, and on <u>social media</u>. We will take a few questions from social media using the hashtag **#AskEHTeu**. An ESO press release will be publicly issued shortly after the start of the conference at 15:07 CEST. Translations of the press release will be available in multiple languages, along with extensive supporting audiovisual material. A total of six major press conferences will be held simultaneously around the globe in Belgium (Brussels, English), Chile (Santiago, Spanish), Shanghai (Mandarin), Japan (Tokyo, Japanese), Taipei (Mandarin), and USA (Washington, D.C., English).

This invitation was sent out for the unveiling of the blurry doughnut that might be a black hole. Unlike the image, the invitation is very impressive, and was clearly based on the templates set out in the 1971 edition of *Nonscience*. Full marks, chaps!

anti-climax of this indistinct blurry shape. The Experts on the stage were roused to a standing ovation, while a few of the audience clapped half-heartedly a few times, until the speaker reminded them of the details: 'What you see here is the result of many, many people working together. This black hole has a diameter of, oh, I forget the number actually.'

As often happens when uninformed members of the Lay Press are present, it was all something of an anti-climax. They realised what they are witnessing is massively important, because that's what they have been told. And they have heard this from an Expert! Had this been a political or economics press conference they'd have jumped on the speaker and demanded explanations—but not in a case like this. What makes this announcement so helpful is that a young astrophysicist named Becky Smethurst took a selfie video of herself behaving manically overexcited by the proceedings. As the expressionless journalists around her lolled in their chairs, arms folded, she filmed herself relentlessly, conjuring up a series of dramatic facial expressions of mystery, delight and joy. She kept posting her unbridled excitement on social media; and the video went around the world. This teaches us all a lesson. The audience won't understand it, but there is no need to

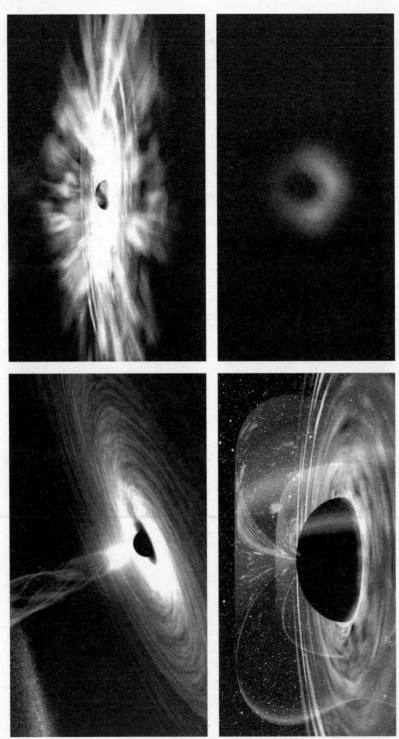

Many images show how astronomers wanted the public to think of black holes—as dramatically detailed whirling fireworks. The example (*top left*) is from NASA, who are celebrated for their integrity; the next image is from MailOnline (*bottom left*), who have been assured that this is what a black hole really looks like; and then the phys.org website presented their own version (*top right*). The actual appearance (*bottom right*) just shows a faint blurry doughnut. It's the worst (and most pointless) photograph ever released.

Here is an object lesson in making any announcement seem fabulously exciting. Young Becky Smethurst takes a video selfie as she sits in the bored audience, to exemplify how excited everyone should behave. These frames show anticipation, surprise, ecstasy, shock, excitement, joy, delight and triumph. Practice these expressions (hint: think of experiencing an orgasm and you should be fine). Then post the result on social media and everyone will realise how excited they're meant to be.

be content with their indifference. Get an Expert to film themselves; and ensure that the result goes public. If anybody ever searches online for the response to the over-hyped announcement, they will see the evidence of excitement for themselves. Practice these facial expressions in a mirror. To get in the mood, think about an orgasm and then overreact. Meanwhile, that black hole image is now part of the history of science. Just don't tell them that many people (including physicist Laura Mersini-Houghton at the University of North Carolina) have proved conclusively that black holes cannot possibly exist. Why spoil the fun?

LONG LIFE MILKED

No matter who you are, everybody wants a long life, so research into this field will always bring handsome rewards. Experts have conclusively proved that you can live longer if you're a fast-walking, tubby, religious, tennis-playing art-lover with a sense of purpose who goes to rock concerts, loves sex, cheese, and tapas, who likes baby-sitting and adores smelling other people's farts. There are sceptics who will remind you that Experts have also shown that it is best to retire young (or alternatively older), that it's healthiest to be happy (or unhappy), both a pessimist and an optimist, to take vitamin supplements or alternatively to avoid them altogether, to drink alcohol (but also to abstain) while getting much more, or alternatively much less, sleep. There is plenty of evidence for these competing claims. No matter—this is all part of the cut-and-thrust of modern Nonscience. People will believe you because of who you are, and they will slavishly follow exactly what you say. To them, you're a high priest of ultimate knowledge.

The public promotion of what you are doing is unrelated to the actual importance of the work. Nonscience replaces religion and works in the same way: unapproachable superior beings whom you cannot fully understand rule your life, and command you to have faith in what they say, whether it makes sense or not. Because we have moved on since religions were launched, the traditional collection plate has been replaced with colossal coffers filled with billions of dollars, but in principle it is the same. Just believe. Never question. And (when asked) pay up. Generously.

Experts pronounce on how to live a long life. Be fat (or thin), a happy optimist (or a sad pessimist). ➡ On page 186, sleep a lot (or a little), retire early (or late), use (or avoid) alcohol; or, on page 187, take (or avoid) vitamins, rave, babysit, and become a Protestant. Page 188 shows you need to drink (or refuse) coffee, walk fast, go to gigs, eat cheese, and smell farts. Is that clear?

Smithsonian.com

Want to Live Longer? Don't Sleep So Much

New research links longer sleep duration with a shorter life

By Erin Blakemore
SMITHSONIAN.COM
MARCH 25, 2015

If you're like most people, you'd kill for a longer night's sleep. But in turn, longer sleep might just kill you. New research shows that adults who sleep more than eight hours a night are at a higher risk of early death.

When Franco Cappuccio of the University of Warwick analyzed 16 sleep studies covering more than a million people, he found that 30 percent of people who slept more than eight hours a night died early, reports Ruth Alexander for the BBC. That's an 18 percent rise in mortality rate from sleepers who reported getting six hours a night or less in the sack.

Why the increased chance of death? Cappuccio, who corrected for depression and sleeping pill use in his review, thinks it has to do with underlying health issues that could be causing longer sleep. But Alexander reports that another researcher has seen increased inflammation and depression in study participants who got an extra two hours of sleep a night—results that could be related to "prolonged inactivity."

THRIVE @ GLOBAL   LOG IN / SIGN UP

The More You Sleep, The Longer You Live: Sleep Science Reveals

"Sleep is one of the most important aspects of our life, health and longevity and yet it is increasingly neglected in twenty-first-century society," Professor Matthew Walker explores.

By Majd Karns, CEO Valu Ventures Inc., Author of THE FIRST DANCER, Board Member @TOWorkforce

In his latest book *Why We Sleep*, leading sleep scientist Professor Matthew Walker, Director of University of California Berkeley's Sleep and Neuroimaging Lab, has provided glaring evidence of why adequate sleep is crucial for the healthy functioning of our brain. Sleep, according to Walker, has transformative powers to change our lives for the better. Yet its importance is exceedingly undermined in modern times, resulting in every major disease in the developed world. Research shows that Alzheimer's, cancer, obesity, and diabetes all have very strong causal links to sleep deficiency.

---

make it   SUCCESS MONEY WORK LIFE VIDEO

You can live longer if you retire early, research shows—here's why

Published Tue, Mar 27 2018·11:54 AM EDT

This couple retired in their 30s and travels full-time in an Airstream trailer

For one, retiring frees you up, allowing you more time to invest in your health. That benefits you whether you're sleeping more, exercising or simply going to the doctor as soon as an issue appears.

Second, work can be stressful, while retirement can alleviate that stress, and stress can create hypertension, a risk factor for various potentially fatal conditions. Retirees in this study were significantly less likely to die from stroke or from cardiovascular diseases.

Harvard Business Review

Latest Magazine Popular Topics Podcasts Video Store The Big Idea Visual Library Read

You're Likely to Live Longer If You Retire After 65

by Nicole Torres
From the October 2016 Issue

RETIREMENT | You're Likely to Live Longer If You Retire After 65

The research: Chenkai Wu, a PhD student in public health at Oregon State University, teamed up with OSU professors Robert Stawski and Michelle Odden and Colorado State's Gwenith Fisher to examine data from the Health and Retirement Study, a longitudinal survey of Americans age 50 and over. When they looked at the sample of 2,956 people who had begun participating in the study in 1992 and retired by 2010, the researchers found that the majority had retired around age 65. But a statistical analysis showed that when people retired at age 66 instead, their mortality rates dropped by 11%.

THE ORANGE COUNTY
REGISTER

SIGN UP FOR NEWSLETTERS
E-EDITION
SUBSCRIBE · SUBSCRIBER SERVICES

Here's why drinking less alcohol can help you live longer

Photo: AP file

By AMY OSMOND COOK | Orange County Register
PUBLISHED: November 26, 2019 at 11:35 am | UPDATED: November 27, 2019 at 7:47 am

Binge drinking is associated with clear risks, but many people m
excess to cause health problems. Heavy drinking carries the hig
even two or three drinks a day could present a problem. Switchi
back a bit could help you lower your risk for these conditions.

Heart disease

Drinking alcohol can adversely affect your cardiovascular health,
more than you should. Alcohol can increase fat levels in your bloo
can also increase your risk of obesity, which in turn increases you
or having a stroke.

Cancer

Alcohol raises your risk of getting cancer, whether you prefer drink
or relaxing with a cold beer at the end of the day. The more alcoho
chances are of getting cancer of the mouth and throat, voice box, e
breast.

LIVESCIENCE

🏠 News   Health   Planet Earth   Strange News   Space & Physics

No, Drinking Alcohol Won't Make You Live Past 90

By Brandon Specktor February 21, 2018

According to the National Institute on Alcohol Abuse and Alcoholism, an estimated 88,000 Americans die from alcohol-related conditions every year,
ited States after
ere attributed

sed risk of
ciety of Clinical
with greater

---

abc 7NEWS   ▶ WATCH VIDEOS   San Francisco

HEALTH & FITNESS

Alcohol, coffee could be key to living longer, study finds

Wednesday, December 26, 2018

People who drink moderate amounts of alcohol or coffee and are overweight in their 70s live longer lives, according to researchers at UC Irvine Institute for Memory Impairments and Neurological Disorders.

The researchers started a study in 2003 to look at what makes people live past 90.

They say participants in their study who drank moderate amounts of alcohol or coffee lived longer than those who abstained from either drink.

In addition, people who were overweight in their 70s lived longer than people who were normal or underweight in their 70s.

EXPRESS
Home of the Daily and Sunday Express

LOGIN

How to live longer: Daily multivitamin could improve health

HEALTH

Global NEWS

Vitamins won't prevent a heart attack or make you live longer: study

BY LESLIE YOUNG · GLOBAL NEWS

Posted May 26, 2018 3:20 pm

Those vitamins you're taking to live longer or prevent heart attacks? They don't work, according to a new study published in the Journal of the American College of Cardiology.

RELATED NEWS

Vitamin D supplements may be useless: Alberta research

Do you even need vitamins? New studies show possible risks; experts offer some safety tips

"What we see is quite emphatically that there isn't a big effect," said Dr. David Jenkins, the study's lead author and a Canada Research Chair in the department of nutritional sciences at the University of Toronto.

In fact, experts say, you're better off spending

NEWS POLITICS U.S. NEWS BUSINESS WORLD TECH & MEDIA OPINION HEALTH SPORTS

Vitamin supplements don't help people live longer, study finds

Dietary supplements not only didn't extend life, but can be harmful at high doses, new study finds.

April 9, 2019, 1:44 AM BST

By Linda Carroll

Dietary supplements don't extend life and might actually shorten it if taken at high levels, researchers reported Monday in the Annals of Internal Medicine. While certain nutrients may contribute to a longer life, they need to come from a food source, the study found.

celebrities, social media influencers and concoctions of vitamins, the new findings join idence that supplements don't help most from the University of Birmingham had found ineral supplements didn't protect against heart

Association of American Universities

Contact Us   LOG IN

Home   People with Religious Affiliations Live Longer, Study Shows

June 13, 2016

People with Religious Affiliations Live Longer, Study Shows

"The study provides persuasive evidence that there is a relationship between religious participation and how long a person lives," said Baldwin Way, co-author of the study and associate professor of psychology at Ohio State.

In addition, the study showed how the effects of religion on longevity might depend in part on the personality and average religiosity of the cities where people live, Way said.

The first study involved 505 obituaries published in the Des Moines Register in January and February 2012. In addition to noting the age and any religious affiliation of the

THE IRISH TIMES

Irish Protestants live longer than Catholics, new research shows

CSO data supports views of historians that Protestants benefit from 'thriftier' lifestyles

Sat, Jun 29, 2019, 00:30

ic Affairs Editor

ased on a data-matching exercise on information taken tics mortality data and the 2016 census on all deaths April 25th, 2016, and April 24th, 2017.

DJ

LATEST   NEWS   FEATURES   MUSIC   TECH   VIDEO   VOTING   TOP-100 LISTS   SHOP

NEWS

RAVING HELPS YOU LIVE LONGER, NEW STUDY CLAIMS

Far from 6AM nonsense this professional opinion is based on heart-rate psychometric tests, which showed participants had increased feelings mental stimulation, and sense of closeness to others while at an event. just 20 minutes of 'gig time' could boost wellbeing by 20%, which is do your average yoga lesson.

"Our research showcases the profound impact gigs have on feelings of happiness and wellbeing with fortnightly or regular attendance being t Fagan. "Combining all of our findings with O2's research, we arrive at a gig a fortnight which could pave the way for almost a decade more ye

Lucis Philippines

Babysitting Your Grandchildren Makes You Live Longer, According to Science

Posted by: Lucis Philippines on December 17, 2019 In Study

Study > Babysitting Your Grandchildren Makes You Live Longer, According to Science

Sure exercise and a healthy diet can add years to one's life — and, evidently, so can babysitting grandchildren. According to recent research, grandparents who watch their grandchildren live another five years on average. When you think about how much exercise it is to run around chasing a toddler, the science isn't all that surprising. But there's more to the science than just keeping physically active — grandparents who babysit are more socially engaged and cognitively stimulated, as well.

The study, published in Evolution and Human Behavior, analyzed data from the Berlin Aging Study, which tracked the health outcomes of more than 500 people ages 70 years and older. In the original main study (which took place between 1990 and 1993), researchers closely examined a core sample of 516 individuals in 14 sessions, covering their mental and physical health, their psychological functioning and their social and economic situation. Since then, surviving participants have been reexamined seven times, as the study has been continued as a longitudinal study.

WORLD HEALTH .NET

LONGEVITY MAGAZINE FREE E-JOURNAL

Articles   Forum   About   you@email.com   SUBSCRIBE

STUDY FINDS SMELLING FARTS MAKES YOU LIVE LONGER

By Health_Freedoms at Nov. 3, 2018, 2:53 a.m., 28328 hits

Pacific Standard

HAVING A SENSE OF PURPOSE HELPS YOU LIVE LONGER

New research shows lower mortality rates for people who feel their life has meaning.

TOM JACOBS · MAY 24, 2019

From the tyranny of ...

NEW YORK POST

LIVING

Study finds fast walkers are 'more likely' to live longer

By Nick McDurden, Tuesday

May 17, 2019 | 2:38pm | Updated

..., of the University of Leicester in England, said: "The findings ... physical fitness is a better indicator of life expectancy than body ...

... lation to engage in brisk walking may add years to their lives."

Rare   ☰ MENU   🔍 f 🐦

New Study Finds People That Eat Cheese Live Longer

Lyndsay Burginger, March 21, 2019 8:08 pm

Sprinkle on another handful of mozzarella on your pizza and grab a bi...
study published in *The Lancet*, claims that eating cheese reduces you...
cardiovascular disease. Now that's something we like to hear.

The study concluded that eating more than two servings of cheese p...

O₂   THE BLUE
News & Views from O2

Published: 27/03/2018 | by O2

Science says gig-going can help you live longer and increases wellbeing

...ncy by NINE years
...gig-time, compared to just 10% for yoga and only 7%
... of live music on the human body and mind

Forbes

Billionaires   Innovation   Leadership   Money

51,034 views | Sep 17, 2018, 07:30am

Want To Live Longer? Take Up Tennis.

Steven Salzberg Contribu...
Healthcare

Tennis might just be the best t...

...ncy rocketing by nine years due to its power to positively ...
... science and Associate Lecturer at Goldsmith's University,
...r Mr. Brightside, and it can be found at your local music ...
... 21% increase in feelings of wellbeing. Added to this,
... with a lifespan increase of nine years pointing to a direct link ...

WebMD   HEALTH A-Z   DRUGS & SUPPLEMENTS   LIVING HEALTHY   FAMILY & PREGNANCY

Heavy Coffee Drinking Tied to Shorter Lives for Younger Adults in Study

WCAI   Local NPR for the Cape, Coast & Islands   Donate

Study: Drinking Coffee Makes You Live Longer

By HEATHER GOLDSTONE & ELSA PARTAN · JUL 9, 2018

A new study by researcher ...
has found that drinking up ...
a reduced risk of early deat...
contradicts this finding. For ...
declared coffee a carcinoge...

So how does coffee really af...

"Like a number of prior studi...
drinking was associated with ...
the 10-year study period," sa...
fellow at the National Cancer ...

SPANISH WAY OF LIFE ›

Why do Spaniards live longer if they smoke and drink? Sex and tapas, says 'The Times'

The British newspaper looks at how the lifestyle in Spain boosts longevity after a study found that its citizens will have the longest life expectancy in the world by 2040

Life expectancy is longer in Spain than it is in the United Kingdom. This is something that has been pointed out by a number of studies, the latest being from the University of Washington's Institute for Health Metrics and Evaluation, published recently in *The Lancet* medical journal. It predicted that by 2040 Spaniards would have the longest life expectancy in the world, and will live on average until they were 85.8 – compared with 83.3, which is the age Brits would be able to look forward to, putting them 23rd on a list of 195 countries.

CHAPTER 10

What the Experts Have Been Doing

At this moment in history, Experts are enjoying a rare degree of freedom of action. Their findings are on every network, in every newspaper; their research is always in the news and their effects are felt everywhere.

Happily it is now possible to bring Nonscience and its powers to the attention of people who, poor souls, might otherwise know absolutely nothing about it. The development of supersonic large aircraft (SSTs in the United States) such as Concorde is an example: even if you live miles away, out in the open countryside, you will hear the reassuring all-powerful boom of the Gentle Giant as she streaks overhead. Not only this, but a number of old churches that were falling into a state of neglect have been resurrected to serve as testing laboratories for the noise pattern (page 20). This is breathing life into old relics—a rebirth that only the Expert has made possible. 'As the old religion dies, the new is born.'

A considerable amount of time and energy has been devoted to the examination of how the Lay Person should think. There have been many different approaches to this problem (more than one being a reiteration of some mental quirk of the author concerned, expanded to see how it fits Humankind in General). One recent thesis is 'Lateral Thinking'. This is a means of encouraging people to envisage alternative explanations lateral to the central problem, so that they get further from it and do not make progress. It also encourages the Lay Person to feel respect for the Expert, who gets to his answer; lateral thinking merely gives a range of impractical alternatives. The writer, Edward de Bono, is relatively Fashionistic right now. It won't last, but with luck the fallout from his books on the subject will confuse the Lay Public (and thereby boost the Expert) for many years to come.

In much the same way but on a far bigger scale, an American Expert, Marshall MacLuhan, has publicised an approach to the media of communication by coining the phrase 'the medium is the message'. This is not bad, as far as it goes. But there is one serious drawback. He uses a very racy and with-it (Fashionistic) style of writing in his books.

Therein lies his downfall. Since his own 'medium' is a light-hearted, frolicking morass of superficial verbiage it follows that his message must be the same. So poor old Marsh is bound to be wrong. Either way he has a loophole. And that is no way for an Expert to behave.

Better thought out is a new explanation of laughter as a Human Phenomenon which has been advanced recently by Dr Charles Gruner, an American professor, and which is a perfect example of Nonscience. Laughter, he states, is a left-over roar of triumph. Many people will be set pondering by this interesting notion. Does it apply, they will wonder, to King Lear begging Cordelia to 'laugh at gilded butterflies'? To the laughter of children on Christmas morning as they behold the presents left overnight by the age-old mystery of Santa Claus? Or the laughter of Pan producing a poet from a mortal man? A belly-laugh over a filthy limerick? But this is hardly the point! The theory has been advanced by a serious and well-qualified Expert who has devoted a great deal of energy and money to the cause, and the least we can do is take it seriously. Impress that on the Mere Layman, if you have the chance.

Desmond Morris, an Expert who used to be in charge of the monkey house at London Zoo, wrote a splendid book called *The Naked Ape*, and another, *The Human Zoo*, which successfully put a host of new interpretations on human behaviour in general. It was a prime example of how you can impose new hypothetical treatments upon outdated notions. Why is it that we assiduously avoid each other as we walk around, being careful to avoid knocking against each other? Desmond's answer was stunningly original, and straight from the school of Nonscience. It is, he explained, because we have to avoid tactile contact as it has sexual implications. The non-Expert, in his traditionalised way, was educated and enlightened by this novel approach, having believed that the reason was entirely different. Most Lay People seemed to imagine that we took this avoiding action because walking is a finely balanced manoeuvre, and knocking into one another would cause people to fall over all the time. But Desmond put us right.

He put kissing in the picture too. Why do you close your eyes when you do it? No doubt most uneducated Lay People believed it was because you cannot focus that close, and you'd only give yourself headaches by trying. But it has deeper reasons when you look at Desmond's theories—and so he went on, through a range of public and private activities. Some Lay People thought his books were 'rubbish'; how imperceptive *can you be*?

Julius Fast, an American Expert, did something similar by writing about non-verbal communication—facial expressions, bodily postures, that kind of thing. It would not have amounted to much (after all, it is a pretty old-hat topic) but he did coin a new term for the subject: Kinesics. And so, whereas you might have been content to believe the employee knocked before entering his employer's office out of deference, habit, courtesy or fright, Julius has been able to delineate this for the non-articulated Kinesic somato-symbolism it obviously really is.

Another American, Edward T. Hall, invented the term Proxemics for his study of distances that separate people. He codified zones of proximity, quantified the separating distances for each, and wrote them up. It produced a whole new study, with hard-and-fast rules, from a lot of ordinary, uninspiring everyday occurrences. Well done there, Ed!

Back home two Experts chased *Capaea nemoralis* snails with an altimeter on Snowdon. Here was healthy, outdoor activity at its most relaxing. Both were museum employees at the time, graduates with higher qualifications, with salaries in the £2,800 region (£42,000 in 2020), plus expenses, of course. The two men—Mr Peter Dance of Cardiff and Mr Adrian Norris of Leeds—chose a low rating for Fashionism in this project. Yet it did very well indeed for them and earned good publicity ratings with consequent professional prestige increments. The interesting aspects of the case are (a) the chasing of snails—itself a relatively infrequent occupation for most Experts—and (b) the use of the altimeter. This is a very expensive item of apparatus which is certainly great fun to play with and a considerable novelty; not only that but its calibration varies with the weather and therefore there's an added opportunity for randomisation in the results, which makes them hard to check and impossible to confirm.

It would have been far more basic to use a map, of course, or to take bearings as indicators of altitude. That is the way the Lay Person's mind works! The abandoning of such elementary and unexciting methods added considerably to the interest of the expedition. Not only that, but it meant that the two Experts were in fact doing their work on the wrong mountain altogether. Full marks for randomisation there, Peter and Adrian.

Applicable as it is to individual cases like this, randomisation is far more dramatic on the grand scale. One such example—the key to Britain's joining that exclusive minority of countries with the same arbitrary standard—is Metrication.

It is a subject which has been born, bred and nurtured out of the womb of Nonscience. The subject of controversy, but the bringer of infinite new benefits to Experts in all walks of life, metrication is a triumph—a pinnacle of the highest order.

The purpose of metrication has been stated as being 'to bring us in line with other countries', but this, as the discerning reader will have instinctively recognised, is only the catchphrase designed for public consumption. Most of the countries we trade with in the Commonwealth do not come into this category, neither does the United States. And even in the European countries that are undeniably metric-orientated there is a tendency to trade on non-metric terms. In France you'll buy flowers by the dozen, in Denmark groceries are often sold in pounds, even in East Europe I've been weighed on machines cali-brated in 'half-kilos', and Italian pasta is packed in libra to this day. In metric Scandinavia, the foot (redefined as 30 cm) is now being revived.

There is a reason for this, of course. Imperial measurements (pounds, inches, and all that kind of old-fashioned stuff) were based on pragmatic amounts that just happened to be self-generating; thus a pound of butter, meat or whatever is a good amount to buy at one time. A pint of milk is a sensibly practical quantity to handle.

But these units, likeable in that sense, are simply not derived by Experts. They are medieval, haphazard relics of a bygone age. It was the French who first decided to make the metre equivalent to an exact fraction of the earth's circumference, which is a real (and not an arbitrary) amount.

And since then we have had a whole range of unitary concepts, all worked out by Experts, all designed specifically by Nonscience; and the old woolliness of the earlier practices has been soundly superseded as a result.

One of the reasons why we use metric units is because it makes it easy to work in tens. Nature does not do this, of course (if anything, she tends to be logarithmic). Natural processes revolve around fractional concepts: cells divide in half, spores are produced in multiples like 32 or 64, and so on. The idea of the figure 10 is itself arbitrary and healthily divorced from the irrationalities of fickle Nature herself. For example, it is impossible to express 'one-third' or 'two-and-two-thirds' in metric terms at all—how's that for getting away from mere pragmatism? And it is obviously more valid to have to express a tradi-tionalised idea like 'a quarter' as 0·250. How sleek, sophisticated and modern it looks to the eye. Fractions are altogether too easy. Thus film-makers knew that, if there are 16 frames in one second, each took 1/16 second—but the decimal equivalent of 0·0624 would soon have added the touch of Nonscience. At any rate, this is what will happen in the future.

While we in Britain were buying our milk in pints, over in Scandinavia they had bottles containing litres, 0·50 litres, 0·250 litres, and so on: a far more contemporary way of doing things.

One fortuitous aspect of metrication is the use of the decimal point itself. It is handy in several ways:

- FIRST, most laymen can't see it anyway, which shows to them how useful it is to have an Expert around to explain these things; and
- SECOND, because abroad the comma is used instead. This makes for splen-did randomisation in practice. Thus an amount such as 29·387 mg would become 29,387 mg on the Continent.

Thus a weight of one-thousandth of an ounce or thereabouts (forgive the use of such an outdated terminology) suddenly becomes nearly thirty-thousand milligrams instead. Clearly this form of phenomenon is going to make Experts more and more indispensable as time goes by.

While we are on the subject we ought to emphasise how metrication is bound to add to accuracy in everyday life (unless the all-important decimal point gets out of position, of course). For example: in the old days one would have said loosely that half the married people in Britain are female; but using the decimalised notion (either 50·0 percent or 0·50 alone) immediately makes one look for the accurate quantification that this implies. The figure looks better, to begin with, and it is firmer and more definitive anyway. It's all a question of degrees of accuracy, which we considered on page 134.

The supplanting of fractions by modern decimals will all add to the mystery of teaching at school level too. Most children find concepts like 'a quarter' as far too easy to grasp—no mental effort that encourages training towards Expertism is involved. For them to think in 'fifths' is too basic when they could usefully grapple with 0·2 (or 0·200, which is of course exactly the same thing when you get used to it). And how many of them have been shown how to obey instructions implicitly by using equipment such as the swimming pool in North Wales which was marked:

3 metre (deep end)
·7 metre (shallow end)?

Experience is the teacher there.

Actually, one really very promising aspect of metrication has been accepted by the government in Britain. And that is the use of the basic unit *the millimetre.*

This fundamental unit, in the interests of Nonscience, is too small to be clearly seen at all by most adults without spectacles or a hand lens (both of which the Expert will always have handy, of course). And greater lengths will add up to be prodigiously complicated figures that only a real Expert can possibly claim to grasp. But metrication was launched upon the public without any serious parliamentary discussion, with no prior public debate at all to speak of, purely as a result of the pressure from the Nonscience lobby. It shows how secure Experts are, in the modern world. Only they could possibly hope to change such a profoundly ingrained system as Imperial measurement into something as arbitrary as metrication.

And the cost at that time? It's hard to be sure, but let us settle for £500,000,000 (in 2020 that's £7,500,000,000, over seven billion pounds) and use it as our baseline. (That is an estimate which several independent observers have mentioned, so it must be right.) This is the cost of, say, 400 new hospitals or over a million homes—all public money diverted toward the employment of Experts and the use of Nonscience.

The stimulus to the economy will go deeper than that, too; all new house designs, cars, machinery and equipment of all kinds will be to different standards. That will encourage new designs for a start.

Why, not long ago I went to a timber merchants and matched up a section of skirting board that was 60 years old with an identical sample in production today. When metrication has fully taken over in just a few years' time, that will be impossible. Instead I would have had to be content with new specifications—to rip out the old skirting and put in a new one would have been the answer. If I need a new tile for the bathroom I will find the metric version doesn't fit. Result? A completely re-tiled bathroom! What a marvellous stimulus to the economy, to the brain, to the aesthetic sensibilities of humankind. We will all have new homes before we know it—and *all* because of Nonscience!

The allied subject of decimalisation comes into much the same category. Not only is the point confused between Britain and the rest of Europe (*vide supra*) but decimalisation in the English monetary system introduced a range of new alternatives. The correct recommended standard is:

£2.56 for typewritten copy;
£2·56 for printed matter;
£2-56 for some cheques;
£2=56 for the other cheques;
£2^{56} for Americans to understand easily; and
£2,56 for our European neighbours.

So that should clear up any misunderstanding. However, we don't want to make decimalisation look too easy, do we? To add a touch of randomisation, therefore, though we called 15th February 1971 'D-day' the changeover in business circles actually began on 1st January—some six weeks early. And the changeover was deliberately stretched over 18 months, during which time either old or new currency could be used in converted or unconverted shops.

One added delight is given below:

(a) Prior to decimalisation, one English penny exactly equalled one US cent.
(b) There were exactly 2·40 dollars to the pound sterling.
(c) 2·40 old pence are exactly equivalent to 1 new penny.

What do you make of these interesting coincidences?*

One thing we must, sadly, admit. The range of choices available for the writing of money values is considerable, and is a triumph of Randomisationality.

---

* 'Absolutely nothing' is the correct answer.

But in the broader, international field of Mathematical Decimalisationalism the range is more limited.

It is still enough to confuse the uninitiated, though.*

We have seen that in Europe the comma is used to separate the units from the tenths thus:

123,456

and we know that in conventional British notation that would mean:

123,456

which is 1000 times bigger. The trouble is that abroad you may find the point (.) used instead to separate the thousands from the hundreds thus:

123.456

which in England would be numerically equivalent to:

123,456

and—again (see how the consistency creeps in here?)—1000 times greater than:

123.456

But suppose you wish to *list numbers in sequence*—what can you do then? In writing you would simply put down:

one hundred and twenty-three, four hundred and fifty-six

but if you did this in figures you would at once give yourself away since:

123,456

means (as the Expert will now readily grasp) *at least three* entirely different things.

In writing large numbers it is now Western practice to write the figures in groups of three thus:

123 *but*
123 456

---

* The initiated, too, if they aren't careful.

which is exactly the form of expression used by some unknowing folk to delineate listed numbers as described in the previous paragraph. And so we can see that the three conventions:

123,456
123.456
123 456

mean separately *eight different things*—and the variations possible if we permutate them all are infinite. The Expert, however, will know what to do.

In 1978 a Frenchman named LeBlond (who also started using a comma after the thousands) suggested that the decimal point be the semi-colon:

123;456

but this fell into disrepute. It could still be revived, if the need arose, of course.

The decimalisation of currency dates back farther, to the Coinage Act passed by the US Congress in 1793, which introduced the dollar (and its 100 cents) as the official currency. It was in the following year that the French followed suit, and now even Britain has done so.

Strictly speaking, a division into 100 is not 'decimal' at all, but 'centimal'. Fractions, on the other hand, are based on normal mathematical expressions—e.g., 1/120 is 10 times less than 1/12. In this way we can see that:

(a) fractions are decimal, really, *but*
(b) decimals are not.

Thank goodness that's all clear at last.

But why stop here? Is it not time to recognise that the movement could be carried further? We have decimalised the money: what of the clock, the calendar and the arc of a circle?

The first decimal-type date was noted down on an Indian inscription around the year 595—and after the French revolution some attempts were made to have a Decimal Day, with 10 hours in one day, 100 minutes in each hour and 100 seconds to the minute. It was sound stuff, Omnipotentially Infallibalistic to the last; but in 10 years it had been forgotten about (no Fashionism, you see). They tried a Decimal Calendar too, with the year of the Revolution as year 1. But that lacked Fashionism too, and (rightly) fell into disuse.

But that is what we need now. There are many combinations (perhaps we could have 10 days per month, or even divide the month into 100 equal parts). Perhaps the best idea would be to have a straight division of the year into 100 days, so that each new day (*n*d) is equivalent to 3·65 *o*d. That would have certain interesting consequences in practice.

In the first place each succeeding day would begin roughly in the opposite part of the diurnal cycle to the previous one, i.e., Monday would begin at midnight, say, and Tuesday would start on *old*-Thursday after lunch. The new day's start would alternate between light and dark, more or less, and no doubt the General Lay Person would complain bitterly about that fact.

But Lay People grumble about almost anything, no matter how trivial. They have put up with leap years for centuries. And now we have begun to get them used to getting dressed in pitch blackness throughout the winter months (by the adoption in Britain of BST*), so the new calendar should be only slightly strange.

They would soon get used to it.

The use of the Decimal Year will mean an increase in the length of the Working Week (since 7 *n*d will be equivalent to 25·45 *o*d) but holidays—at 50·9 *o*d—will be wonderfully relaxing. Even if we took the fortnight's holiday to be 0·5 *n*-month (by analogy with today's practice) it would still be 18·25 *o*d, an increase of 4·25 *o*d on present standards.

The naming of the days of the week would be simpler, too; thus instead of MONDAY and TUESDAY we would be saying ONEDAY and TWODAY.† The months would go well with the new notation, too.

We would start the year with

ONEUARY,
TWOUARY, and so on right down to
OCTOBER,
NOVEMBER and
DECEMBER

(which, since *Octo*, *Novem* and *Decem* are Latin for eight, nine and ten, are ready-made for our Decimal Year and could always be translated into English if the need arose).

We then can look to an increased spread of this terminology until it pervades our entire way of life—and just think of the progressive benefits it brings!

---

* We justified this change by saying that it was 'impractical' to keep changing from Summer Time to Greenwich Mean Time every year and, happily, this has been generally accepted. The fact that the United States manages with its time zones was ignored in this argument, as was the continued use over there of Daylight Saving Time. And so we had the near-ultimate in Progress—where Greenwich Time was used the world over (except by the inhabitants of Greenwich). A pity it didn't last.

† The abbreviated form of Oneday, Twoday, etc., would be written '1d, 2d'—thereby reviving an older convention for our earlier (non-decimal) currency system.

It would work as follows:

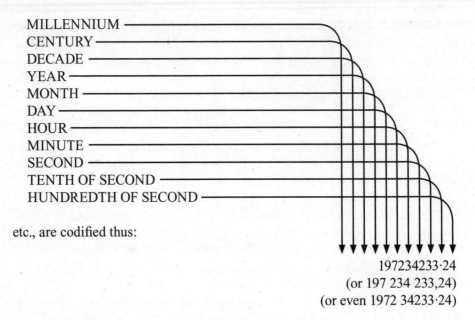

MILLENNIUM

CENTURY

DECADE

YEAR

MONTH

DAY

HOUR

MINUTE

SECOND

TENTH OF SECOND

HUNDREDTH OF SECOND

etc., are codified thus:

197234233·24
(or 197 234 233,24)
(or even 1972 34233·24)

and see how much simpler it would be in practice:

OLD TALK:
'The time is now just gone half-past one on the third of April in the year one
thousand nine hundred and eighty three *anno domini*. I weigh exactly ten
stone at the moment, and I stand just five feet and seven inches from head to
toe in my stockinged feet ...'

NEW TALK:
'At 198334593·40 I am 63·5040 kg & 1701·8 mm.'

And why stop here? In October 1971 the Japanese set out with a scheme (only
a *pilot* scheme, at present) to identify the inhabitants of major towns and cities
by numbers instead of names. As an experiment, 14-digit code numbers are to
be used so that the ancient custom of personal nomenclature (Miss Softness of
a Lotus Petal; Mr Might of the Enraged Tiger at Dusk) can be replaced by arith-
metically composed data-codes instead: 29837465134786 or 93847091157362,
for example.

Meanwhile we have zipcodes, postal codings, all-figure telephone numbers
and so on (not to mention National Health Numbers, Family Allowance Book
numbers and the rest) which have been willingly, almost gratefully, accepted
by Ordinary People.

THE SECRET OF THE STAMP

Nobody has yet realised that the inscription on the new British postage stamps is another example of Nonscience propaganda. The Lay Person thinks that the printed symbols mean '½p', but they clearly don't:

(a) The symbol for a new penny is 'p': on the stamp it is printed 'P'—the CAPITAL letter.
(b) The amount is believed to be '½', but the '1' is actually a capital letter 'I' when you look closely.
(c) The correct decimal way to write 'half penny' is £0·005 anyway, and not ½p at all.

So, since the figure '1' is really a letter 'I' and the 'p' is actually 'P', what do these strange initials signify? Only this: it is an indication of the subtlety of Nonscience, for IP stands for Implicated Professionalism. Experts calculate the degree of Expertistical Professionism, multiply it by the success with which it is Implied to the Public (i.e., I × P) and then divide by two to obtain the mean. That's why I P appears on the stamp.

So we can now confidently look forward to the day when we can abandon the kind of woolly nonsense we have inherited from our long-forgotten forebears:

'... and my name is George Quentin Fuzzbody the Second, of two hundred and ninety-seven Ash Street, Midville; just go to the supermarket on the corner and bear to the left for half a mile or so ... we go to that pleasant doctor by the bus terminal, you know the one?'

in favour of a condensed, complete rundown in figures. That way the complete NEW TALK designation would look like this:

'I, 94837615564783 of 149 E44 St Zip 993424A FA: LG 6670394NHS No WRCL 9724, NR No: 14522/a/PB(90329)29b at 198334593'40 am 63·5040 g & 1701·8 mm.'

Crisp, concise, meaningful. An economy of words yet a super-abundance of information—the programmer's dream.

THE EXPERT IN HISTORY

With all this feverish activity we tend to forget about the solid background to it all that Experts have laid in the past. Robert Hooke in the 1660s obtained a microscope and with it drew quite a number of specimens as seen with its aid. That was a very popular pastime, as it happens—Sir Christopher Wren had a microscope, so did Samuel Pepys and most of the smart young men-about-town of London at the time boasted their own instrument too—and Hooke's book *Micrographia* was a sure-fire bestseller. His was the exploitation of Fashionism at its best, and he added to the book several sections of his own ideas just to get them in print.

It was, of course, a success. But his views on light (included in his book) were completely ignored until they were expounded and elaborated by Newton some years later, even though they were probably the only theoretical work of any real consequence that Hooke ever did. Why? The answer is simple— Fashionism again. The microscope was Fashionistic when Hooke published *Micrographia*, and Newton found that light was Fashionistic by the time he launched his own publications on the subject some years later.

Much the same applied to Pasteur, of course. Lou—another peerless leader of our movement—was a pioneer of the press conference. He would hold demonstrations in fields in the best manner of an open day (page 174) and invite all the right press people along to watch. He became famous for his Fashionistic exploitation of bacteria and fungi, much of it work which had actually been done in the 18th century, long before he was even thought of, by an obscure Italian monk. But it took Lou to present the material in a popular manner and win support, and to him belong the honours. Oddly enough, his own interesting work (he was in his spare time a scientist, though it should not be held against his memory) on crystals and optical isomerisation was never widely publicised. But then, that wasn't Fashionistic either.

Charles Darwin used much the same kind of ploy too, by writing his thesis at exactly its most Fashionistic time, when everyone was discussing it. He wasn't the first to propose his particular interpretation, of course, but his use of Fashionism and the clothing of the argument in detailed observations of animals in general made the whole project an obvious winner.

The earlier pioneers of antibiotics found much the same comments applied to them and their failure to carry popular favour. In fact, even when Fleming discovered penicillin it didn't get very far. But he was lucky, for a dozen years after the discovery that *he* made (following on the heels of all the others who had done it before) the whole matter suddenly became Fashionistic and, before you could gasp with shock, there was Alec in the headlines at last. Penicillin—and its Expert—had arrived!

There was Freud too (discussed later in this book as an Inspirational Figure on whom one can base one's Life, page 241) who transferred his own mental abnormalities into a set of theories which he applied to people in general, and Marx who wrote about the Science of Society from the standpoint of someone who was not even peripherally involved with it in any way. His arguments may lack thoroughness, in that he based his entire work on the publications of others, and not by doing anything original himself; and it is perhaps devoid of that authority which he might have gained if he actually knew anything about Bourgeois Principles or Human Nature. But he arrived on the scene at a perfectly Fashionistic moment and deserves his fame as an Expert well enough.

And we cannot leave Edward Jenner out of our list either, famous as the discoverer of vaccination. Actually he was nothing of the sort, of course. His Fellowship of the Royal Society, *q.v.*, was granted because of his amazing discovery that it is the young cuckoo that pushes the other eggs out of the nest. Sadly this wasn't Fashionistic, and it has therefore been forgotten about. Vaccination hadn't been Fashionistic either, when Rabaud-Pommier had investigated it. But Jenner knew it was a topic of conversation when he carried out his famous trials on a half-wit clergyman's son, and with this boost of Fashionism, he knew, he was bound to become famous. It worked, as it always does.

Great names all: Bob, Lou, Charles, Alec, Ziggy, Karl and Ted, we salute you!

What a sad spectacle alongside these men are the so-called 'scientists' who busied themselves making new discoveries in a non-Fashionistic vacuum.

There was Cesalpino in the mid-1500s classifying plants by floral structures as we now do; Vesalius, founder of anatomy; and Nicholas of Cusa who in the early 15th century knew more about planetary motion than Copernicus, who became the Expert in that field when it was Fashionistic a century later. There was Peter Peregrinus, who wrote of magnetic poles in the 13th century long before it was Fashionistic, foolish man; and Viete, founder of modern algebra in the 16th century. No-one today has heard of Römer, propounder of light's finite velocity, or of Hall and his achromatic lens of 1729 (it is Dollond in the 1760s who became the Expert in this field).

Malpighi in 1660 proved the theory of circulation of the blood; and shortly afterwards Hooke realised that fossils were extinct species—a most important advance in its way, but it was not Fashionistic in the least and therefore he is remembered for entirely different activities (page 200). Guettard founded geology in the mid-18th century, and back in 1682 Blaes (known to his friends as Blasius) had written the first book on comparative anatomy. Lamarck discovered the idea of evolution before 1800.

Buffon, back around 1750, had first mooted that kind of idea, but it wasn't until 1852 that Spencer really brought it out into the open. Russel Wallace thought of 'natural selection' at the same time as his friend Charles Darwin, but it was Charles who published his *The Origin of Species by Means of Natural Selection* which took the bookstalls by storm and established him as the world's Evolutionary Expert for all time.

Back in the 18th century too we find Haller, who was the first man to publish on the information explosion and who tried to systematise bibliographies (page 135). But no-one remembers him either—and who can name any of the countless Gloucestershire folk who knew about vaccination before Jenner was even born? Hales discovered the science of quantitative physiology in the early part of the 18th century, and Sylvius 50 years before had already set the seal on a chemical approach to life itself. But that wasn't Fashionistic either, at the time.

Taking the patient's pulse was not practised until the 19th century, though it had been investigated by Galileo (known for work closer to Fashionism) and Sanctorius in earlier centuries—and that was another of the ideas first recorded by Nicholas of Cusa, once again.

In the field of chemistry, Scheele was busy discovering gases and important chemicals at the same time as Priestley, but he lacked the latter's presentation. It was Fashionistic, all right; but where are you without the factors alluded to on page 203? Nowhere, in a word. Atomic theory had trudged along its way too, over the centuries; it was proposed by Bede in the year 600; by Isidore, the Spaniard, around 700; and was popularised by Gassendi early in the 1600s. But by the time that Dalton came along, though he was but a north-country handloom weaver by trade, it was a subject of Fashionism—and it is he we remember with respect.

There are so many forgotten men in the field of electrical research too. It was in the 1660s that Guericke made the first continuous electric current generator, in 1729 that Gray discovered conductivity through wires, and 10 years later Kinnersley (one of the very few American scientists in history), at the same time as Hauksbee and Gray, was discovering two kinds of electricity (positive and negative) and showing how an electrical charge rests on the outside of bodies.*

Davy, at the beginning of the 19th century, discovered a host of new observations about elements (several of which he isolated for the first time); gases, acids and bases and all that kind of stuff. None of this had Fashionism at the

---

* In the Apollo 12 moonshot, which was struck by lightning shortly after blast-off, it was this very phenomenon that saved the craft from destruction. Of course, the charge was *bound* to remain on the outer surface of the rocket, for the reason discovered back in the 1730s. But it was far more newsworthy to discuss what *would have happened if* the charge had passed through the inside of the spacecraft instead. So that theoretical drawback was carefully avoided, thereby making the whole event seem *far* more exciting.

time, and so it has all been forgotten about. But his miner's lamp was launched at the peak of Fashionistic popularity—and everyone remembers *that* to this day.

The mapping of the brain—by men such as Bell, Magendie, Legallois, Hall, Flourens, Broca and the rest—was done in a low-Fashionism era too, and so we can recall none of these men.

We remember the Expert Wheatstone who measured the speed of light in 1834—but we do not recall that it was the same order of velocity found by our old non-Fashionistic friend Römer and his contemporary Bradley over a century before, do we?

And who has heard of Carnot, Mayer, Seguin, Roget and Colding, who founded the modern view of energy and its conservation? Though we do not know it, Carnot's work pervades the whole of our modern practice in that field. But he was not Fashionistic and inevitably went down, poor chap.

Fraunhofer discovered spectral lines in 1814, successful (if minor) Expert that he was; though admittedly the same work was pioneered by Wollaston in 1802 and a Scot named Melvil back in 1752. But there was no Fashionism for the topic then either. It was in 1797 that Cruikshank discovered the mammalian ovum, just when the subject was not Fashionistic. By the time von Baer came back to it in 1827 the climate of opinion was altogether better—and with Fashionism on his side he obviously stood a far better chance.

Xavier Bichat, founder of histology in the 18th century, has been the victim of low-Fashionistic placing in history too. So have Lobatschevsky and Boylai, who were writing on curved space a century before Einstein. There are countless others in the same category, needless to say.

But why did these antiquated scientists fail to become Experts? One of the most important reasons was a Fashionism failure—a low *qu*FN rating. As we have observed repeatedly, it is *not* necessary to make a new or original discovery. That may well hamper your activities as much as anything, and the responsibilities that follow will certainly tend to cramp your style.

Far more important than this is the necessity to be Fashionistic—so *do* remember that. It's not all that counts, though. Some of the examples we have seen were perfectly *qu*FN-orientated but lacked the grasp of the media so necessary for success. There is no virtue in being an Expert only in the compass of your own small department, remember; get before the public eye—that's what matters. And to hell with the empty, mind-contracting concepts of 'originality', 'discovery' and the rest.

Even in today's technological era, when we like to think that we have the names of innovators firmly established in the public consciousness, we find the same tendency. Names such as Marconi, Baird, Fleming, Diesel … these have become household words. But it is only because of their *adherence to the principles we have outlined*. Many others fell by the wayside because of a failure to uphold the principles of Nonscience. Usually it was a Fashionism lack. Occasionally it was an inability to appear suitably omnipotent or gifted; sometimes because of a deficiency in using the mass media effectively.*

Fleming and penicillin are well-enough known. But who bothers to show interest in the pre-Fashionistic observations of Gratia Dath and Rosenthal in 1925 (years before Fleming's *Staphylococcus* culture was ever contaminated by the stray spore of *Penicillium*) that living bacteria could be dissolved by an organism called *Streptothrix*—a discovery that led to streptomycin? No right-minded person recalls the names of Carothers, discoverer of Nylon or Whinfield and Dixon, inventors of Terylene; all pre-Fashionists for their subject. The name of Sundback, who developed the zip fastener, is pretty well known in the right circles; but the man who actually invented it, back in the 19th century, has been totally forgotten by all but historians. His name was Judson.

Although it is a vitally important process, xerography was actually discovered—as most things inevitably must be—in a relatively non-Fashionistic climate for the subject and so the name of Chester Carlson is not known either. Nor is Peter Goldmark, inventor of the long-playing record; and Valdemar Poulsen, who invented tape recording at the end of the 19th century. Dr Normann perfected margarine in 1902, though he too never hit the headlines. By the time that became interesting he had moved on to a process that has given us synthetic detergents—but they were still pre-Fashionistic at that time, so he lost out there too.

The name of Baird—a peerless Expert whom we should all revere—is, and rightly, a household word. He will be forever associated with television, as securely as Darwin with survival of the fittest and Pasteur with spontaneous generation. And yet, like these other great men, he was not the actual lone pioneer. Indeed he was, from the historical viewpoint, an almost totally insignificant figure.

Cathode ray tubes were first commercially available in 1897, and the possibilities for their use as a picture-transmitting system—television—were recognised by Rosing in 1907 and, in considerably greater detail, by Campbell-Swinton over the following five years. By 1912 it was realised that the simple mechanical disc form of TV transmission was clearly going to be useless for long-term development, and from then on two inventors, Zworykin and the self-educated Farnsworth, pressed ahead with perfecting the system we have today. Neither can be classed as Experts in our sense of the word. They were mere inventors.

But Baird it was who took the simple rotating disc system and brought the full resources of his most *Ere*N-orientated personality to bear. He could effectively sell himself to the media and thus, though his contribution to television

---

* Factors known to Experts as *Qu*FN, SS, IO and E*re*N, you will recall.

is in the limited sense of value, quite worthless; nonetheless it is he who rightly deserves Expert status today.

So we can see that there are many more unsung inventors, forgotten pioneers and overlooked experimenters than there are famed Experts of world renown!

Why did these lesser individuals fail? The reasons are clear to the student of Nonscience. Often they failed to utilise the potentialities of Supracharacteristic Sociotropism (selective snobbery, as it was once unkindly dubbed); frequently they did not sell their ideas through the media as they should have done—in fact their grasp of elementary press conference procedure is often lamentable—and underlying many of the failures was a simple, cardinal lack of Fashionism. And that, as we have seen, they could have rectified by their own efforts. But, above all, these earlier workers had one fundamental drawback. It is of peerless importance, towering above one's SS and IO ratings, exceeding even the stature of Fashionism; *without it you are nowhere at all.*

The factor they missed was the sleek, well-bound, informative volume you are reading at this moment. Without the concise, considered evaluation of Nonscience of which you now have the use, cutting laserlike* to the core of the problem, they clearly could not see the wood for the trees.

You, dear reader, are not liable to that drawback. With this treatise available at all times you cannot but succeed.

MYTHS AND LEGENDS

It is surprising how little some things have changed. When *Nonscience* was first published in 1971, metrication was being introduced. It didn't happen—in Britain we still use both systems in a muddled manner that confuses everyone; everybody except Experts. Beer is sold in pints, but we buy petrol in litres, and measure fuel consumption in miles per gallon. Carpets are still made in rolls 12 feet wide, but are measured in square meters, and priced in square yards. The millimetre remains the standard unit of length in Britain and is the only measurement standard in the world that cannot be clearly seen by anyone over 40. We are stuck in a halfway house where nobody wants to settle for one system or the other. Indeed, with Brexit only a memory, there are proposals to bring back pounds and ounces. Bushels and pecks can only be a short way behind. There is nobody available to establish a permanent system of weights and measures.

Experts do not innovate. They indulge in high-cost, repetitive activity. Indeed, since Fashionism is the bedrock on which the subject is based, innovative ideas

---

* The older term would have been 'razorlike' of course. But that was in the days before lasers were Fashionistic.

can only spoil your status. You are never a trend-setter! People were surprised at the curious stories of innovation I published in the original *Nonscience*, yet it is interesting to reflect on how often the amateur spearheads discovery. The people who launched the modern era of science were the first to employ the microscope, Leeuwenhoek and Hooke. Leeuwenhoek was a draper, Hooke an assistant with no training in microscopy. Christopher Wren, famous for his stupendous buildings (like St Paul's Cathedral) was no architect, but a physician who designed buildings for a hobby. Herschel, who discovered Uranus, was a chapel organist (he was assisted by his sister Caroline, who drew up pioneering star catalogues and discovered galaxies and comets, but was largely ignored since she was a mere woman). William Smith, who compiled the original six-foot-long geological map of Britain, was a navvy; Robert Brown (of Brownian Motion fame), who recognised the cell nucleus, was a doctor; electricity pioneer Michael Faraday was a bookbinder; Charles Babbage (who, with Ada Lovelace, designed the first computer) wanted to work out the odds when betting on horse races; and Charles Darwin failed his University studies (he went to medical school in Edinburgh, but was rusticated, and later did a degree in divinity) and was the captain's travelling companion aboard *The Beagle*, and not the official naturalist. Pasteur was no microbiologist, but was an industrial chemist who once worked at Whitbread's Brewery in London; Mendel (who discovered genes) was a priest; Dunlop, of pneumatic tyre fame, was a veterinarian; Strowger, who invented automatic phone dialling, was an undertaker eager to stop his rivals eavesdropping on his calls from clients; Carlson, who invented photocopying, was a lawyer keen to cut down on the amount of copy typing done in his office; and the Wright brothers, aircraft pioneers, were originally bicycle makers searching for a more profitable way to use their steel tubes.

Crick and Watson did not discover DNA! Crick (just like Carl Sagan) proposed that life was seeded on earth by super-intelligent spacemen, while Watson originally studied ornithology and featured on the *Quiz Kids* show on television. Their achievement was working out the structure of DNA, a field in which neither of them had experience, using data from Wilkins and Franklin. They never did any research on DNA, nor used it in any of their experiments; indeed, they had been forbidden from interfering in that investigation. Even today, most innovators are unqualified amateurs. Steve Jobs (founder of Apple) had no qualifications above high school. Nor did Bill Gates, who dropped out of Harvard aged 20, or Sir Clive Sinclair, who left high school to become a writer, or Mark Zuckerberg, who didn't complete a degree when Facebook became a success.

There is a clear trend here: unqualified amateurs lead innovation, not Experts. It is almost a rule of progress—the great steps forward are taken by gifted rebels, while the Experts trot along behind, claiming vast sums of money to repetitiously refine what an ingenious individual has founded. So, who was the greatest and most innovative scientist of the 20th century? Would it apply to them? The accolade surely goes to Albert Einstein, hailed the world over as a superb innovator.

Einstein, who did poorly in most subjects at school, was working as a patents clerk in Zurich when he published his revolutionary theory of relativity. It revolutionised science. Yet it was done, with nothing more than a pencil and notebook, by a clerk, and in his spare time. Such is the true nature of innovation.

HISTORY MYSTERY

What we believe is what we have been taught, and not what is true. History of Science courses (like the latest one at Oxford University) perpetuate pointless fictions through the generations. They proclaim how the greatest mind of the 17th century was, of course, Isaac Newton. What they skate round was his stealing of ideas from people like Robert Hooke (whose portrait Newton destroyed, to stop Hooke becoming famous), his belief that you could convert base metals into gold, Newton's prediction of the date of the Rapture, based on the Bible, and his recipe for a concoction that kept you alive for eternity. Newton wrote about spirits and ghosts, astrology and alchemy. Like many other Experts, Newton was nuts. No wonder he featured as a member of 'the Priory of Sion' in *The Da Vinci Code*. Hooke made dozens of important discoveries, from the nature of colour (which Newton stole) to the design of the escapement mechanism (without which traditional clocks would never have existed). Still nobody learns about Hooke, though they all celebrate Newton. You don't learn about Leeuwenhoek either, even though he gave us our first insights into living cells and microbes, which matter to us all today. The historian's view of 19th-century science celebrates Charles Darwin above all, even though the theory of evolution had been published long before his time. Darwin's book on worms greatly outsold his book on evolution during his lifetime.

People know of Alfred Wegener as the reviled individual who introduced the idea of continental drift* before the First World War, after seeing how the east and west coasts of the Atlantic Ocean fit together like a jigsaw. Not so. That was first discussed by Ortelius in 1596 and noted by Bacon in 1620. Antonio Snider-Pellegrini suggested in 1858 these continents had once been joined, and by 1891 Alfred Russel Wallace (the man who originally wrote 'Darwin's' theory of evolution) was writing that continental drift was a commonplace idea among geologists. A few years later, William Henry Pickering wrote that the continents had once been joined together, Eduard Suess proposed all the continents had once been a single landmass that he named Gondwanaland, followed by Frank Bursley Taylor who proposed the theory to a meeting of the Geological Society

---

* Continental drift was soon abolished by Experts, who insisted we call it Plate Tectonics instead. Continental Drift explains precisely what's happening, whereas Plate Tectonics doesn't mean a thing to the public.

This is a clear picture of continental drift. I know what you're thinking: 'Aha, a diagram from Alfred Wegener, who published the idea in 1915!' Not so. Back in 1858, an investigator of whom you've never heard, Antonio Snider-Pellegrini, published the theory in *La Création et ses Mystères Dévoilés* (Creation and its Mysteries Unveiled). But he was an innovator, and didn't follow any of the Rules of Nonscience. Of course he was ignored, and quite right too.

of America in 1908. By the time Wegener came along the theory was widely understood throughout Europe, though not America. He called it 'continental displacement' and since then it has been renamed 'plate tectonics' because those earlier names were self-evident and tectonics didn't mean anything to the public.

Scholars don't know this, and so the encyclopaedias, newspaper features, television documentaries, schoolbooks, websites, and magazine articles all state that Wegener had the idea first. Like Darwin and his theories, scholars always get it wrong. Plate tectonics is an everyday subject now, but not until the 1970s was the theory thought respectable by American geologists. See how long it takes for a new idea to become accepted? Stick to the old ones. That's the only assured way to fame, money, and power.

CHAPTER 11

The Way Ahead

One question remains: What are the most propitious avenues for you to explore? A considerable amount of care must be given to the choice of career if fame and fortune are to be yours. The decision to enter a low-*qu*FN sphere of activity—one that was not Fashionistic, in other words—would put an end to such ambitions almost before they had begun. Finding a job when you come out of college (page 80) is one thing: finding a niche from which to rise to power may be *quite* another!

There is nothing much doing in pure chemistry these days. New manufacturing procedures and so forth are always available if you are hard up for a job, but there is nothing very Expert about that and you'll never catch the attention of the headline-writers at that rate. Plastics are *passé* too, even carbon fibres have fallen into disrepute just when the RB211 jet engine looked as though it was going to make them a household word.

But in the biochemical field there are more promising outlets, particularly when we get into the realm of pharmaceuticals. We have gone through hormones and steroid drugs now, and even antibiotics are un-Fashionistic these days. Let that not deter you! It is highly acceptable just now to be in the forefront of drug research, largely since pressures are being brought to bear that will undoubtedly lead to the legalisation of marijuana within the next few years. Current betting would put Denmark and Holland neck and neck in the lead of the race towards positive legislation, but there are states in the United States (New York, California later perhaps) that could easily follow suit.

The main argument is to equate marijuana with tobacco and alcohol. Actually, in the spirit of our Movement, this is not strictly true when looked at from the point of view of factual data (i.e., the old 'scientific' approach).

The main distinction is that alcohol and tobacco are both widely taken—by the greatest number of users by far—purely for the flavour of the concoctions themselves. Many people like whisky because of its warming, soothing, honeyed flavour. Businessmen smoke cigars because the scent of their smoke is pleasant, and the aroma that lingers in a room is somehow manly and delicate (perhaps this is a modern equivalent of the burning of incense, Experts!). Pot, however, is used *solely* for its mind-altering propensities. Thus, in spite of the popular view, it comes into an entirely different category from these other legalised materials.

BUT DO NOT EVER MENTION THIS FACT. This point of view has never yet appeared in the literature, and you must not attempt to promulgate it. The fashionable view holds that marijuana is acceptable, that legalisation is the coming trend; you court disaster by attempting to launch any alternative viewpoint. Neither, come to that, must you be influenced by personal feelings about the matter. Pot is here, it's here to stay, and a vast new industrial complex (taking over the resources of many of the tobacco companies) is ready, waiting in the wings for its cue. So get on to the bandwagon promptly, but above all do it with fervour.

The law on the subject shows the degree of uncertainty which surrounds it. Thus in the States you can be gaoled for 33 years for possessing a single reefer in Texas; it would be a few dollars fine in New York; and you'd probably be ignored in Berkeley. In Holland the local municipality opened a youth club in Amsterdam, the Paradiso, where soft drugs are openly taken.

So, if you've tried smoking pot whilst at college, you could take it up into a whole career, and be in the forefront of progressive campaigning as a result.

Just at the moment there's a Fashionistic new kick for the medical biochemist that would make a very prestigious and profitable career. That is the study of *prostaglandins*. There is one disadvantage: results from the research are probably going to be wanted by an employer. But there is an advantage that somewhat offsets this: results are dramatically easy to come by, and almost anything can—at this stage—be claimed. Without going into irrelevant detail that really belongs more in the dead era of science than in its modern counterpart, the prostaglandins are a series of fundamental hormones with far-reaching properties. Just at present they're being postulated as cures in the following fields:

- asthma;
- sterility;
- for the abolition of the symptoms of the common cold (which the Expert should *always* refer to as coryza); and
- as contraceptives. They are already being used, clinically, as post-coital abortifacients which are inserted, intravaginally, if the lady doesn't menstruate on the appointed day.

But what about ingrowing toenails, habit tics and spasms, bed-wetting, menopausal tension, hayfever* and all the other scourges of humankind? For as long

---

* Cancer, leukaemia and multiple sclerosis ought to be on the list too, if it's to reflect accurately the modern tendency.

as the prostaglandin craze lasts there is no limit to the number of possible breakthroughs that the Expert can postulate.

A more traditional aspect of the same subject is the code-naming of 'newly discovered' drugs.

In this the Expert takes some well-known product and attaches a code name to it—XL99, Buggex-32 or whatever—and from that moment it can be marketed at a very much increased profit margin. An example is the selling of aspirin/phenacetin/caffeine preparations which, as APC, sell at around 15p (in 2020, £2.25) per hundred, but with a brand name instead (there are dozens in current use) fetch up to £1 (£15 in 2020).

The health food boom is a very successful outlet for enterprise of this sort. Health preparations cleverly suggest that they 'do not contain chemicals'. Naturally, *we* all know that they are entirely composed of chemicals (like all foodstuffs in existence), but is there any need to bother the Lay Person with these incidental data? True Experts know that it is their duty to do the thinking *for* the public, and not trouble the unwittingly ignorant minds of merely Ordinary People with subjects they do not grasp. So, if people feel happier eating food 'which contains no chemicals', then let's produce it.

It is quite simple. One example, just to set you on the right lines, would be to emphasise that your product contains such health-giving materials as Vitamins C and B_6, instead of harmful ascorbic acid and rotten old pyridoxine. One manufacturer of beetroot juice claims that it benefits the blood (because it is the same kind of colour) and, although strictly speaking this claim could be said to be without any possible foundation in fact and therefore, in that sense, untrue, many people buy the product and are no doubt satisfied to imbibe the end-products of the Expert mind.

Sometimes government agencies foolishly work against the natural interests of Nonscience. Recently in the United States the Federal Food and Drug Administration authorities have condemned a number of leading products as being 'ineffective'. The leading toothpaste, made by Colgate-Palmolive, was one of them. It contained a material known as '*Gardol*', one of the trade-named products of the type we have mentioned above. The answer to this form of needless meddling is to word advertising claims with care. Thus one should not state that LIBID-*o* 'increases sex drive', but that it 'can help you to enjoy sex as never before'. Not only is the latter choice less definitive, and suggests only a possibility anyway, but it sounds more subtly exciting.

But in medicine generally there's always *something* newsworthy to do. Let's look at how the ordinary, run-of-the-mill statistics of an ordinary, run-of-the-mill Medical Officer of Health can be used to advantage.

Dr David Morgan, MoH for Chester, noticed during his 1970 totting up of the figures that there were 13 cases of neonatal deformity in his area *more than in the previous year*. What does this mean, out of the teeming thousands under his care? Frankly, not a great deal; it's hardly a statistical novelty. Even if it was, there was nothing to connect it to anything else, such as a change of dietary habits or the use of some mutagenic material in the area's industry.

Ten years earlier the radioactive fallout problem was much in the news, and had he been active then in this capacity it's likely that Dr Morgan would have seen in this the possibility of a successful gambit—'it's the bomb,' he could have said.

But this was 1970, remember, when that particular subject had dissolved away into the vapours of time and un-Fashionism. A very highly Fashionistic topic had taken its place—the Contraceptive Pill. The Pill is certainly the most extensively monitored material ever ingested by anyone, and in spite of the considerable benefits it brings, there are only the slightest hazards attached to its use—hazards that fade into total insignificance if compared with the other facets of our modern life. It has been calculated, for instance, that the safest way for any woman to be is on a maintenance routine of the Pill (since the risks of any other form of contraceptive and the associated hazards of mortality are considerably higher). But nonetheless there has grown up a formidable anti-Pill movement and the whole subject has become so Fashionistic that one can *guarantee*—positively—that any statement about any aspect of the Pill (particularly aspects that tend to make it look dangerous) is bound to come in for attention by the media. It is intensely Fashionistic—and why not?

A great deal of down-to-earth, honest employment has been given to Experts the world over in the anti-Pill campaign—Experts who would otherwise have been literally out of work if the Pill had simply been *accepted* without question. And the anti-Pill industry had ensured that the media were kept fully primed with facts and opinions to maintain public interest. Dr Morgan, quick to realise this, at once made a demand for a public enquiry as he published his figures for the year. He was quite sure, he stated, that the Pill ought to be investigated as the possible culprit. There was no actual reason to assume that there was any connection between the Pill and the figures he was reporting, he added; but only a national enquiry would find out the answer. So he went down as a figure of importance, and an Expert of some standing.

In the general way, being a Medical Expert can be very tiring. But it can be very well-paid too, and it is extremely glamorous. The nicest thing about it is that you are completely safe. When the lives of Ordinary People are at stake the penalties are not severe, even assuming they are not covered by professional insurance. A London Medical Expert in 1968, for instance, neglected a patient and did not bother to send him to hospital when he probably needed it to survive. He was fined 50 guineas (in 2020 that's about £800). In 1970 the same Expert declined to attend a pregnant Lay Person who later became seriously ill, aborted the foetus and underwent an emergency operation. But here his fine was only 150 guineas (£2,400 in 2020) which is cautionary, though

DO YOU NOTICE ANYTHING?

SYMPTOMS ASSOCIATED WITH:

Taking the Pill	(b) Taking a placebo (i.e., dummy pill)	(c) Chronic worry
HEADACHE	ACHING HEAD	HEADACHE
DEPRESSION	FEELING DEPRESSED	DEPRESSIVE STATE
BREAST TENDERNESS	TENDER BREASTS	A FEELING OF TENDERNESS IN THE BREASTAL REGIONS
NAUSEA	FEELING SICK	FEELING NAUSEA
VOMITING	BEING SICK	SICKNESS
WEIGHT GAIN	INCREASE IN WEIGHT	GETTING HEAVIER
NERVOUSNESS	BEING NERVY	NERVOUSNESS
TUMMY ACHE	STOMACH ACHE	BELLY ACHE

There is a marked similarity between the symptoms of the three conditions when you look *really closely*. No-one has realised that many of the side-effects attributed to taking the oral contraceptive are actually due to worry about the publicity of harmful side-effects. Many Experts have made a packet out of (a) commenting on the risks, and others have been just as happily employed in (b) treating the results of the worry that follows. But now Dr J. W. Goldzieher of Texas has carried out 'controlled trials' which show that Pill effects occur in women who are given a pill which is actually only a harmless substitute made of inert materials.

This shows, according to him, what a lot of money, time and printed paper has been wasted in the past. Wasted, Dr Goldzieher? Diverted to the ends of Nonscience, if you please.

not crippling. Take the case of the provincial hospital doctor a couple of years back who was carrying out a very fascinating investigation of a patient's brain ventricles. The patient was not actually suffering from a brain disease, and the aim of the investigation was to ascertain some properties of the healthy brain. Some sensationalising scandal-mongers would no doubt call this an 'unnecessary human experiment'; but it is, as the sympathetic reader would no doubt agree, merely a part of the Expert's stock-in-trade to find out such things. And

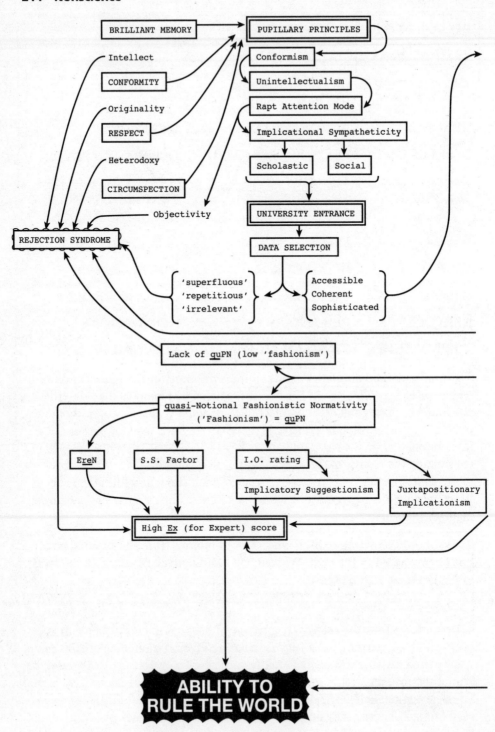

The master flowchart for success, fame, fortune—and power.

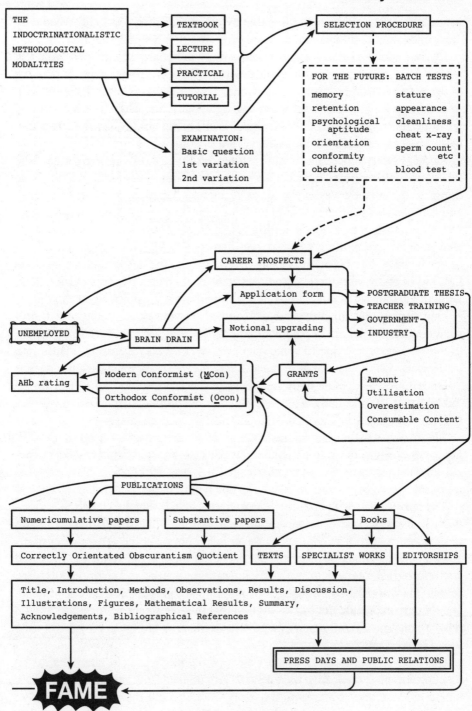

why make a patient tense and irritable by asking for their 'consent' when they don't understand anyway and will only fret unduly? It simply isn't on.

During the manoeuvres the telephone switchboard summoned him (by means of a breast pocket bleeping device, which is always very impressive to the patients) and he went to the phone. The caller, as is often the case with impetuous young ladies, kept him hanging about longer than he expected and when he returned the patient was dead. Well, better to die like that than in a road accident, as we say in Nonscience; but of course—in spite of the advances in medical Nonscience that could result from this kind of investigation—the relatives would no doubt have been emotional and obstreperous about the whole thing. So the episode was explained as being a 'collapse on the operating table', and everyone was satisfied. No point in raising issues unnecessarily, after all.

But suppose the patient's family had heard? There is a good statistical chance of this happening, when we consider the weekly prevalence of this kind of occurrence. The problem is easily overcome, as the defence of the case, its fighting in court, and the eventual payment of damages (if any accrue) along with all legal costs, is covered by medical insurance. In short—you can't lose.

One survey into this very subject showed that out of a sample of 270 preventable deaths of women patients, patients and Lay People were to blame for 91, and the rest were due to some oversight or (probably excusable) lapse into carelessness on the part of an Expert. There were no repercussions, of course.

Once in a while, busy-body members of the Lay Public like to call for an 'ombudsman' to take care of the investigation of complaints in cases like this—but fear not; that won't happen.

The matter was put before the BMA in 1970, after it was raised in an NHS green paper. One committee member, Dr Alex Skene, said 'This will be a tribunal … where doctors would be challenged'—a dreadful notion. Naturally, the proposal was rejected, unanimously, as one would hope.

The only trouble with being an Expert in this field is that the Health Service—itself an example of Nonscience at its most prolific—has unfortunately begun to develop in such a way as to inconvenience not only the fringe of customers who expect that sort of treatment, but even the staff themselves. Cramped conditions, insanitary quarters, dirty operating theatres and suchlike are often found. They are the consequences of randomisation when it gets out of hand, and though one would not like to suggest that the Nonscience Movement should rebel against proliferations, it can be a little irksome at first until you get used to it. Devotees soon do this, of course.

But many Experts prefer to steer clear of the medical field altogether. As a result some new operating theatres are left unused and empty, which pours a healthy douche of water onto those arguments about the supposed 'lack' of such facilities. For instance, in August 1964 a female patient was reported to have died of gangrene contracted after an operation in an old, neglected theatre of a northern hospital while there were two new theatres alongside, unused through staff shortages. That shows the scope available for Expert employment, if naught else. And of course it was in October 1964 that a Birmingham eye theatre had infected six patients, five of whom lost the sight of an eye or two, and the other of whom died; so there is plenty of interesting work waiting. Modern outbreaks of things such as hepatitis tend to infect staff as well, mind you, so it would be as well to be duly cautious before rushing in willy-nilly. Lay People we can lose, but Experts have to be replaced. Remember that.

Do be careful to avoid too much scandal if it can be helped. In one recent case an Expert at his routine coffee break (and let's face it, one must stop sometimes) had a young patient suffocate. If we have to lose them that's one thing; but this way of doing it—though likely to supply information that many Experts would like to have access to—can cause questions to be asked. That is never wise.

In Britain the respect for Medical Experts is profound. A case is on record of a man who attended a hospital with a cough, was misunderstood, and found himself undressed, on a trolley, and confronted with operation consent forms. He was (quite rightly, one might add) so imbued with respect for the Experts around him—whom he instantly recognised as being his betters—that he signed the forms for major surgery and was wheeled into the theatre. It was only then that the facts came to light. Many other patients, no doubt, have gone right through the mill without ever noticing anything was wrong, having had operations and all. But it is better to be *too* respectful and deferential, than not enough.

In recent years some members of the public have complained about treatments that cause disease, instead of curing it. The concept of iatrogenic disease has arisen as a result.*

Iatrogenesis is a good term since it

(a) cannot be easily understood by the non-Expert; and
(b) has a pleasant, lilting ring about it that diverts any suspicions harmlessly away.

For example, it is now known that monoamine oxidase inhibitor drugs (known in the trade as MAOIs) can have a most drastic and dangerous reaction with foods containing pressor amines. They include beer, cheese, bananas and foods like that. Now some people have collapsed and died after eating a cheese sandwich while they were taking the tablets. Needless to say, in this world of

---

* Actually the term means 'doctor-producing', and so the only truly 'iatrogenic' agencies are the parents of students who eventually become doctors. I do hope that's clear now.

CONDITIONS IN SEARCH OF A CAUSE (ANY CAUSE)

Cancer
Multiple sclerosis*
Hiccups
Strokes
Ulcers
Fainting fits
Sterility
Heavy rainfall
Low sunshine figures
Ingrowing toenails
Earthquakes
Anaemia
Disseminated sclerosis*
Spina bifida (*note: for pronunciation aid see page 82*)
Schizophrenia
Hot flushes
Cot death syndrome
Genetic malformations in general
The common cold outbreak
Malnutrition (Western countries only)
Premenstrual tension
Warts
Acute disseminated encephalomyelitis*
Nystagmus
Urinary incontinence
Sclerotic demyelinisation syndrome*
Vertigo
Bad breath (halitosis)
Hairy ears
Leukaemia
Coronary heart disease
Baldness
Runny nose
Prickly heat
Eczema
Epilepsy
Cyclones (general)
Alcoholism/drug addiction/smoking
Alzheimer's disease of the elderly
Palpitations
Militant nationalism

* A rose by any other name …

distorted values, it is the poor Expert who gets the blame. But one must ask this: is it right that the Nonscience Movement should accept blame for everything? Can we be expected to run around and check what people are eating? They should be expected to find these things out, and not be so apathetic as to allow them to be left as unasked questions. The legal system in this country states that 'ignorance of the law is no defence' and quite right too. We should recognise that the same applies to Nonscience. People who want to take pills should take the trouble to find out about them, and not wait for tragedy to strike before they complain.*

It is in the operating theatre that the most eye-catching developments generally take place. And one must recognise that Medical Experts, if they are to attain any degree of stature, must follow the basic rules of Nonscience, this means *action*. Take organ transplantation. There have been many research workers in both the globe's hemispheres who have carried out endless experiments with animals. Now that is all very well if all you wish to do is work out a new operative procedure. But true Experts, the people who are really going to go places, do more than this.

They get *on* with it and have the Press in—quickly. Patients are going to be lost, of course, as is inevitable when experiments are under way. The unsolved rejection problem means that the patient's long-term prospects are not good. But is this important, when Nonscience is at stake?

Whether you are interested in Lay People or not, the field of medicine is a gratifying one in which to be an Expert. One can tell this from the relationship that exists between the Medical Expert and the other types—specialists in biology, chemistry and so on—with whom they work. Medical Experts look down on them. The degree of implied superiority varies, but it is always, invariably, to a greater or lesser degree, there. For this reason non-medics would be advised to steer clear of the medical laboratory unless they are unaffected by this kind of atmosphere. It is all part and parcel of the discipline, and it is essential in a field such as this to have a healthy social gap between the Expert and their subject. It would be unthinkable for patients to begin to question all that was done to them—one could not have that kind of thing going on.

The financial side is good. Salaries of over £8,000 (£120,000 in 2020) are quite feasible, with added perks from textbook writing and suchlike. In addition the chances for favourable treatment from traffic wardens, policemen, theatrical seat agencies and so on are very favourable.

Without doubt, forensic science is another very jolly and interesting subject. It seems to attract many people with a somewhat macabre bent, naturally enough; and they write of their work in a quite characteristic fashion. Even in

---

* Or their dependants do, if the patient succumbs to the effects.

the most tragic and horrific cases, terms such as 'a remarkably good example' are found.

There is plenty of scope for humour, too; one standard textbook features a photograph of a murdered woman smothered in blood lying in pools of congealing fluid with lacerations over much of the upper part of the body. The corpse is thrust back, hanging over a lavatory pan and the face is cut and disfigured. The handle of a sheath knife is seen protruding from the chest. And the caption of this scene? 'Fig 2: Injuries arousing suspicion' ...

Yet forensic specialists are just that—true specialists. They do not concern themselves with the work of others, mainly because they have other work to do. Thus their texts do not as a rule have anything to say on the methods of reviving the near-dead victim; they merely advise specialists to *make sure that the person is dead* before starting their work.

There are large amounts of money to be made from this occupation. An appearance for the defence in a trial can be handsomely rewarded. There are many examples. Not long ago a wealthy man was blown to pieces in an explosion as he got into his car. Traces of odd fragments of wire were found, but much of the evidence had melted away (as was, no doubt, intended) in the ensuing blaze. Nonetheless, there was clear evidence of foul play, and one of the suspects was an Expert in explosives. But, as the case was drawn together by the deft arguments in court, a forensic Expert was brought in by the defence. He asserted that a build-up of petrol vapour under the bonnet could have caused the blast. *Could* have, mind you. There were ample figures available to suggest that the force which could have been generated in such circumstances was nowhere near enough to result in the damage seen, but that—in a court of law—is immaterial. The important thing is that the Expert has said that the alternative explanation *could* have applied—and that is just the kind of *reasonable doubt*, as it is called, with which one cannot convict. The fee for such a small service is very large (it can run into five figures) even though the amount of time involved is very small.

The trick of Forensic Nonscience is to pick out whatever evidence you need to substantiate your particular point of view.* Thus if your client wishes to be seen as the innocent member of a motoring dispute, you might end your dissertation on an accident by saying: 'It was tragic that a peaceful afternoon excursion could have been brought to such a tragic end by means I have outlined.' Or if you have been brought in to discredit *the same individual* you might say: 'It would seem appropriate to bear in mind the evidence of recklessness to which I have alluded earlier since, though one hesitates to apportion blame absolutely, it is clear that without this form of thoughtless—one might say dangerous—behaviour, accidents of this kind would be very much rarer.'

---

* This is the Margument in action once again, of course (page 107).

The important thing to realise is that *nothing has actually been said*. The first (defensive) explanation is a self-evident truism, and says nothing that might actually apply to the facts of the case. The same is true of the second (accusative) explanation. It is obvious that accidents would be rarer without dangerous driving, and the fact that one can 'allude to evidence' in one's learned exposition does not mean for even a moment that anything of the sort *actually happened*. The best thing about being an Expert is that you can always get away with bland statements of inference which counsel can return to later in a more definite manner.

Let us briefly consider an example: a hair found on a cardigan, for instance. The two approaches are:

(a) THE DEFENSIVE
The Expert in this situation puts the disclosure of the hair into its proper context by arguing:

    i. that the hair was of typical appearance;
    ii. many people have hair of this kind;
    iii. the variation of hair on any head is considerable;
    iv. the defendant was in contact with many people from whom such a hair
        could have originated … and so on.

The role of counsel is then to allude to these arguments retrospectively and say: 'We have heard Expert evidence which clearly and unequivocally shows that the presence of a hair on the defendant's cardigan is neither here nor there. There is no reason to convince us it had anything to do with the victim. We all have stray hairs on our clothing, and one could not draw sinister implications from this! No doubt many people in this courtroom have similar hairs about them at this moment, but would we claim that as evidence? Of course not. It is even possible that the wanted man is himself here today: but it is abundantly clear that he is *not* in the dock.'

(b) THE ACCUSATIVE
In being called for the prosecution or the plaintiff the Expert's task is somewhat different. They would, in this position, argue:

    i. that the hair had been found on the defendant's clothing;
    ii. that another hair taken from the victim had been compared with it;
    iii. that in all respects the two were similar and
    iv. that a common origin was compatible with this evidence.

The duty of counsel then is to add the touch of conviction thus: 'And finally there is the tell-tale hair on the cardigan worn by the defendant on the very night of the crime. In size, colour, shape—in short, in all the visible respects that are apparent to the searching eye of the Expert in his laboratory; with all the evidence of modern scientific knowledge—this hair was utterly identical with the victim's. One may, I suggest, draw one's own conclusions.'

In sampling hairs, incidentally, it pays to hunt around a little. Whether you wish to find a single hair similar to the specimen, or one that is quite different, you should not have to search for long.

But no matter if it is fragments of moss, particles of mud or dust or any other kind of microscopic evidence, it is an easy matter *either* to produce a totally dissimilar sample from the same source in order to show that they could not have had a common origin; OR find a similar sample in virtually any situation to suggest that the evidence is sound. I remember a classic piece of evidence used in court on one occasion was the superimposition of a skull on the full-face photograph of the victim before she was missing. The two were made to fit exactly, which of course can be done with any skull at all within reason. The court accepted that evidence. It would have been just as easy, of course, to prove the opposite.

That leads us to a postscript; a rule-of-thumb for the Expert who enters this field. Proof is not necessary. All you need is a statement that the evidence is 'compatible' with so-and-so; or that in your 'considered opinion' it is most unlikely to have done so; counsel can harden it up into evidence. The jury (and the judge, too, come to that) won't understand you anyway. Probably the best thing about being a Forensic Expert in Britain is that there isn't any national Forensic Service. Anyone can be called in at any time, and so the scope for original interpretation is considerable, and the chances of being argued with are slight. The Expert will get on well with the legal profession anyway, which makes this kind of activity even more socially enjoyable. The modern legal business has a great deal in common with Applied Nonscience and its aims are similar, too. So you should feel quite at home. There are hundreds of successful murders committed each year which are never detected, and opportunities on both the defence and prosecution sides of lesser offences are legion. A doctor does not even have to see a dead body to register it as being dead, remember; which all serves to add to the fun.

The related subject of microbiology is a field crammed with possibilities. Micro-organisms are found everywhere, and they have practical implications that are so diverse and varied that no-one could ever hope to unearth it all.

There is therefore ample scope for research and for rapid progress too.

For instance, a whole new type of organism has recently been unearthed, the mycoplasmas. (Actually they were first recorded scores of years ago but only recently have they become really Fashionistic; so everyone *thinks* they are new discoveries and it is best not to disillusion them.) They can be found virtually anywhere in any warm-blooded host, and almost all stages of the life cycle seem to be transmissible. Think of the scope here! They could be the subject for headline-catching research papers of almost unlimited breadth. You could find them in association with arthritis, cancer, mental disease; and suggest that perhaps this is the cause at last. You could cast aspersions on almost any social convention as a risk involving the possible transmission of the organisms (kissing, shaking hands, talking, public transport or anything that takes the fancy).

All micro-organisms produce metabolic end-products, of course. Don't we all? And these can be produced in very large amounts if we culture enough of the organism. Penicillin is one example, alcohol is another, dirt-dissolving enzymes are a third. You could find organisms could produce literally tons of other things—such as uric acid, paint pigments, drugs of all shapes and sizes—with enough diligence. And, what's more, you could very easily find species (or mixed cultures, now *there's* a clue) that could render garbage, waste oil, plastic refuse or whatever into harmless soluble residues. Quite a lot of interest has recently begun to centre on these problems, and it could well be a useful field to follow up. Fame and fortune lie in wait for the person who patents the eventual remedy for waste paper and refuse. And it is abundantly clear that the organisms are available, awaiting discovery; waste paper, leaves and so on quickly rot down in the ground. But there would be no need to divulge the exact nature of the discovery. That would only make it seem basic and unexciting.

In the human field it is quite possible that you might find populations of bacteria and fungi in the intestinal tract respond to changes in bodily behaviour, so that an alteration in faecal odour—as one example—could indicate the presence of a symptomatic change elsewhere in the body. Or you could explain a whole range of bacterial conditions in psychogenic terms if you were crafty, and that would be a sensation.

Perhaps the whole psychology/micro-organisms/smell field is going to break through to Fashionism quite soon. It started with Groddeck (one of Freud's students) who insisted that we all had a sense of smell as good as any dog's, but it was suppressed because of psychosexuasociological reasons. Recent research has shown that there just might be something in 'odour therapy', and didn't many psychologists in days gone by insist they could smell a schizophrenic?

The new Fashionistic term for such substances is pheromones (the first in literature was written about by Dr Alex Comfort; he called it 3-blindmycin). You may think that you wash your armpits so that you do not smell offensive to a loved one—but with this new idea in mind couldn't it be something else? Are there not long-term effects of deodorants; indeed, might not washing prove to upset a whole range of notionally pheromonic human responses?

And if we take the notion further we can easily suggest that it is specific populations of bacteria which cause these smells (in some cases) in the first place: *so they might be catching!* Perhaps there's an odour of being in love;

perhaps you and your proximity to a loved one brings about a transfer of organisms until you share pheromonicalistical odorations (or 'smell the same', as the Lay Person would say). Diet could change the balance slightly and so make you more acceptable to particular people ... and there must be applications to the diagnosis of disease.

There are even possibilities attached to the use of bacteria in the *treatment* of disease. Genetic disorders could be altered around the use of micro-organisms that produced a reverse effect.* The avenues are almost too many to contemplate in detail.

And, of course, it is going to dawn on someone that bacteria are potentially immortal (since they reproduce by fission) and so the chances for learning something about self-regulating non-ageing systems are many. One would only have to start work on that field to be hailed as the 'person who is going to make us immortal'! That would be valuable in television fees alone.

Many micro-organisms reproduce sexually, it is now known, and the chances here are for cross-breeding experimentation. Botulin—a bacterial toxin—is still the most poisonous substance known, and has been for a century. So there is one world title waiting to be challenged.

But investigating the bacterial population of soil, houses and homes, even of ourselves, could be made to throw up all kinds of exciting new relationships between the familiar world and the microscopic universe.

What changes take place when a couple make love, for instance? Does personality affect microflora? It's a fascinating field, pregnant with possibilities.

Viruses are another thing altogether, of course. No-one has a cure for any virus disease—not a reliable cure, at any rate—but research has now come so close to the answer that this would be a very propitious field in which to work. The lucky person can make their discovery world famous in next to no time, and it could well be hailed as a masterstroke in our war against disease. Nothing much need be said about vaccines, which have taken much of the credit already for the *prevention* of these illnesses, or about our degree of uncertainty over the exact significance of viruses. But they have been *suspected* in the causation of several human cancer types, and have been *mentioned* in connection with many other diseases—from arthritis to warts—and so the popular appeal of such a development could be made to seem vast.

The other aspect of viruses, of course, is their destructive nature and the fact that we know so little about them. Thus the field is wide open for anyone to discover a supervirus that could wipe out humankind† or to suggest new

---

* As this book was being compiled, someone did this with a virus. But the field is still wide open.
† Luckily enough, there are no laws controlling dangerous experiments with bacteria and viruses.

viruses as causative agents in other diseases. How about schizophrenia, for example, heart attacks, the kind of phlebitis that causes clots to form (resulting in strokes and coronaries) and the like? These are a few suggestions to start with.

It is worth noting that the virologist can always manage to evade too much criticism. Thus, if a rival insists that it is bacteria that produce a given effect (or mycoplasmas, page 222), the sharp virologist can quickly counter by saying that it could be a virus carried by the parasite that is responsible. And once we get down to the genetic significance of viruses—i.e., the view that they are all escaped genes, or that many are lying latent in our bodies anyway—we can see that the whole subject, once again, is open and inviting.

There is a good deal of international interest in all these problems, and research funds are easy to raise. In particular there is *Brucella*, causative organism of brucellosis (which was once called contagious abortion) in cattle. It can infect humans, too. Nothing much need be known about this organism before you begin research, as no-one knows very much as it is. Similarly, once again, any discovery is likely to prove important. The British government alone is pumping the best part of £5m (in 2020 £75m) per annum into this subject at the time of writing and, since the condition itself costs £2m (£30m in 2020 values) per annum the brucellosis expert would clearly be on to a Good Thing.

Just at the present time we have begun to recognise the presence of 'slow viruses' in the human body. This term has now been accepted to describe a number of entities that exhibit the following features:

(a) no-one has ever seen them produce any repeatable effect;
(b) they have never been observed by the electron microscope or by any other means;
(c) it has never been possible to isolate them in any way;
(d) they cannot be cultured in the laboratory by any known method;
(e) they are apparently unaffected by boiling or by disinfectants;
(f) they do not produce the typical anti-virus reaction of the body;
(g) complement fixation tests (normally used to prove the presence of viruses) do not work with them;

and yet they are there all right. There is very wide scope here for the 'discovery' that 'slow viruses' cause almost any condition you care to name. Within a few years, when the topic has become more Fashionistic, research funds will be proliferating like mad.

It is interesting to see the way that Nonscience has progressed through the pathways of microbiology. Until the beginning of the 20th century, poliomyelitis virus was everywhere. It was a very common childhood infection, where it caused a short-lived feverish illness. The literature on the subject states: 'infants hardly ever became paralysed by it.' As a result, adults were in a state of immunity to the disease.

But by then the hygiene consciousness provoked by Experts such as Pasteur (page 200) swept across Europe and the incidence of the illness fell. Children were no longer being exposed to the live virus, and so by the time the infection *did* come along it took hold in a far more severe fashion. For scores of years they died or were crippled on a vast scale—until we found the best form of vaccine, which is now generally administered before the age of risk.

Thus we are back to exposure of children to a living virus. That is also what used to happen in the Bad Old Days, of course.

Vaccines as a whole are one interesting and fairly easy way of spending a few Expert years. There are many disease organisms for which no vaccine has ever been found, and it's only a matter of sitting around, waiting, while technicians plate out controls and watch for plaques on cell cultures.

But the crux of the microbiological world at the present time lies in the field of genetics. It is not a difficult matter to extract the fundamental biochemical units that control heredity—DNA and RNA—and manipulate them in the laboratory. Some small changes can be made; it is possible to disassemble part of a molecule and put it together again; it has even proved feasible to synthesise certain genetic units *in vitro*. Each time a breakthrough occurs it is heralded by the mass media as being a step towards the creation of life. It isn't, of course; but that's hardly the point.

This is a broad front of progress in which new developments are easy to make and (since the actual creation of synthetic life of any kind is very far away) each small step along the pathway is likely to be greeted with great enthusiasm for years to come.

Just recently, for example, one Expert put the cell membrane from one cell with the nucleus and cytoplasmic mass of two others (all from the same species of *Amoeba*, mark you)—an interesting task, though not particularly novel. But he was at once hailed as the 'creator of a living cell' and the last I heard he was apparently being interviewed about synthetic living systems in general. The truth of the matter is this: the subject falls into two clear categories: synthesis of living matter on the one hand, and alteration of already existing material on the other. Apart from simple grafts of that kind, it would take only a very small genetic change to produce an entirely new form of organism, to create a monstrously deformed experimental animal—or human being, of course.

Once the altered genetic messenger apparatus had been implanted into a living system, nature would do the rest. Now, this also has the appearance of being 'creation', and, carefully used, this ploy can give an aura of prestige and drama to any such development.

Actually, the creation of living material—i.e., the synthesis of an entirely artificial living cell—is quite impossible. We have yet to learn infinitely more than we know at present, and our knowledge of the detailed happenings within the cell—how it divides, to begin with basics, as one example—is extremely limited. If we list the activities of a cell we find that the overwhelming majority cannot be even rudimentarily understood at present. Indeed, many of them cannot even be listed, as we do not yet know what they are ... but is there any reason for Lay People to know this? Such a revelation might only give them feelings of mistrust and insecurity. So we must guard this secret well, and know that our very ignorance itself gives us the power to create forms of life that no person is likely to understand.

But which *everyone* will read about.

Of course, these biological matters have Fashionism on their side. But many other branches of our Movement have lost a great deal of their Fashionistic appeal of late. Space research is in a deep depression just now, but is due for a revival in the 1980s. Nuclear physics is at a low ebb too, though there are opportunities for those who wish to work with subatomic particles. But it isn't newsworthy at the moment. In allied electronic fields the development of electron microscopes is at a peak. Scanning electron microscopes are the latest rage and now every self-respecting institute or department should have one. There is room here for a modern Hooke (page 207) to do an updated *Micrographia* that would sweep the board with record-breaking sales.

But the main branches of physical research remain the computer and lasers. Lasers, only now little more than a decade since their discovery, are big business. Many Lay People like to pretend that Experts have never yet found a successful application for lasers: they suggest that this noble and exciting device is 'a discovery looking for a job to do'.

Nonsense! The laser has found many 'jobs to do', among them:

- hastening the growth rates of radishes;
- serving as a stage prop in the British Cockpit Theatre;
- co-starring in a James Bond movie;
- as a means of ensuring 70% target obliteration in Vietnam bombing raids;
- acting as a cheap nuclear warhead detonator;

and several others which escape me for the present.

Just to take one Expert application of the laser principle, we will refer to the newly developed Lidar, announced by the Oak Ridge National Laboratory in the United States. This is a laser assembly fitted with a detector which feeds its signal output into an oscilloscope. The equipment is being used to look for smoke coming out of the 800-ft chimneys of America's biggest and brightest new power stations.

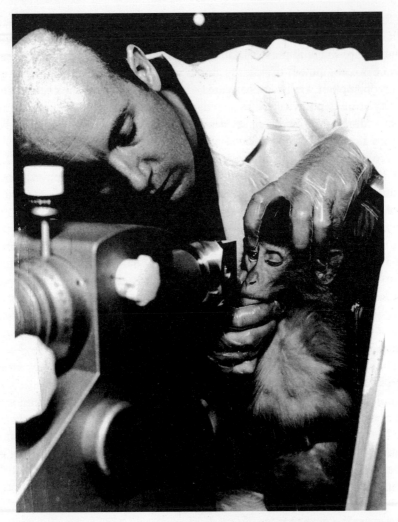

The discovery of lasers suddenly revealed, for the very *first time in history*, that Experts did not know what happened when a laser beam was shone directly into the unprotected eye. Might it not, for example, cure blindness? This Californian experiment will soon throw some monochromatic light on the subject.

The centre of the equipment is a pulsed gas laser which has the usual optical exciting apparatus. The monochromatic beam is directed towards the chimney smoke and reflected light, picked up by the detector, is converted into an electrical potential charge by the amplification unit and thence fed into the oscilloscope. The trace on the screen indicates whether or not the chimney is smoking. Thus this monitoring facility is entirely automatic and totally orientated towards technological complexity. It is a fine example of what can be done today.

In the old days, when there weren't such things as Lidars, people had to detect smoke in other ways (like looking).

Then we have computers, a most gratifying branch of Expertism. The very presence of a computer* in an installation adds a fundamental note of authority. 'Using a computer, Dr Smith has found ...' and 'The latest computer forecast says ...' are examples of the formulae that are recommended for use. Any sketchily drawn-up notion, once processed through the machine, acquires a heady aura of solidity and trustworthiness which is as convincing to the Expert as it is to the ordinary, run-of-the-mill Lay Person.

Of course, the use of computers has long been greeted with catcalls of derision by the anti-Nonscience movement. Here we have an excellent means of rendering specialist data incomprehensible to Lay Persons (who would be far too brainless to know what to do with it, and have to be protected from themselves) and, what is more, of maintaining vast banks of information ready for Expert use whenever necessary.

The Antis have at once christened this trend the 'data-bank society' and it has become very fashionable to decry the whole trend. Just look what damage to the progress of Nonscience is resulting from all this!

The General Register Office, in conjunction with the Royal College of General Practitioners, launched a survey in 1970 which is a case in point. The details of British patients were to be filed away and categorised in a central computer store, including such information as whether or not the individual was on the contraceptive pill, if they had had venereal disease, what were their personal medical quirks and so forth; along with names and addresses, of course. When the first notifications to GPs were sent out it was made quite clear that in carrying out this survey, the consent of patients should not be asked for in case of repercussions. Indeed, the doctors were advised that they should not tell their patients anything about the use of their private records in this way, 'since many would refuse and the survey would not then be fully representative.'

This warning, however, was not heeded by some hot-headed young doctor whose identity will never be known. The news was leaked to the press, who with the wilful mindlessness so typical of Lay People immediately started an 'outcry' against the whole thing. It was an invasion of privacy, they said; it was wrong to divulge confidential details of a personal nature, and so on; the arguments came thick and fast. As a result, a perfectly sensible and progressive piece of Nonscience was jeopardised. This all goes to show that, where Nonscience is concerned, the less people know about the ins and outs, the better. They don't understand (how could they be expected to, without the years of conditioning and mind training it takes to become an Expert?) and they will only cause problems.

---

* I take it you have one?

Secrecy is the by-word.

At this juncture it is worth pointing out that the proliferation of data-storing facilities gives the Expert some clear advantages. The mainstay of such systems is the punched card, and ordinary members of the working classes (the masses, that is to say) are mystified by the little rectangular perforations that randomly decorate their monthly credit statements, invoice checks and so forth. Experts know better. They are conversant with the system in use and are readily able to read back and decipher such documentation. One hates to mention it, but several underhand and thoroughly nasty individuals have even misappropriated this Expert knowledge in the perpetration of forgery and fraud.

Such a system used by a defecting Expert filched £50,000 (£750,000 in 2020) out of one of Britain's largest caterers, and an American speculator was able to change his bank's overdraft limit from $200 to $200,000 with the aid of a razor blade and a little care. Perhaps the future will bring a vast expansion of such criminal activities, which may well provide a source of interest for disinherited Experts (both aiding the operation and helping in its detection—both, if you're careful).

Certain Enemies of the Movement even see, in such simple means of interfering with computer records, an easy way to sabotage the 'databank society' altogether. So be warned, Experts, and *keep your knowledge to yourselves.*

Our colleagues in the United States have taken some firm steps in the right direction, and have not been deterred by pettifogging trivialities thrown up by the unthinking public. The American Defense Department has a computerised databank covering 25,000,000 citizens. (Incidentally, don't be free with this information. It is supposed to be kept quiet.) Each day around 12,000 requests for data are processed, and 20,000 new items of information are added. The

The standard IBM type punched card code used to interpret (and sometimes amend) credit statements, invoice cards, etc.

accuracy of the bank will obviously fall as time goes by—but why worry? It is at least a start.

Of course, it is secrecy that matters, and in industry that becomes a prime factor. Millions of pounds, countless dollars, can be at stake. Recently there have been examples of rival computers demanding information from secret memory stores held by rivals—and getting it, too. That poses some exciting questions for Experts at large, and is bound to ensure even more new avenues for employment in the future.

The ultimate state of society as exemplified by Nonscience is, as the Expert knows well, when all the trivialities of life can be cut out by the general use of computers. The health survey may have been blocked by 'public pressure' (actually the lobbying of Interested Factions, if the truth be told) but the government already has taken the first steps towards the systematic encoding and listing of all personal details by computer. The record cards the doctors keep are government property; so are the little files they are kept in. They can be recalled at any time for checking. And we have seen the public acceptance of telephone tapping which is surely a good omen for what is shortly in store. Listening-in by police on private telephone conversations is permitted in the national interest—but details, figures, instances or an indication of reasons for doing so are not available since that *isn't*.

So we can be sure that, when it does come, the Common People will accept the computer as a friend, and as being in the interests of the state as a whole. Then, incomes will be automatically taxable from computer records, people's personal details can be available to government agencies, and the whole tenor of life can be improved as a result. No longer will the aged be lost to the world, the poor starve in garrets, without us knowing about it.

We will know *all* the details—and all because of Nonscience.

Meanwhile, the gathering of information that Ordinary People don't know about is gaining speed. Social security dossiers contain material that one really couldn't disclose to the people they were written about, and many patients suffering from cancer have been listed on national cancer registers to further research. One can always say in a case like this that it is impossible to seek permission from the individuals, since most of them don't know they are in that condition anyway. So the creation of the principle is already accomplished.

Errors in computer programs have often been cited as one reason to steer clear of them. But this is one of their most delightful characteristics. The media have for too long proclaimed that computers are 'dangerous', or 'inhuman'; that they 'depersonalise' our way of life. But when some pensioner gets a gas bill for £750,000, a moon rocket flies off into space because of a misplaced comma, or we sit with our hearts in our mouths (as happened in Apollo 11) because a near-tragedy results from a computer error; surely we can see how futile this objection becomes. Computers can perpetrate mistakes like the rest of us! They are not so 'inhuman' after all! Should we not articulate those wise words:

TO ERR IS HUMAN, TO FORGIVE DIVINE

and be happy that no-one is perfect?*

There is another very satisfying field of employment as an Expert in design—that is, the turning of new developments into actually marketable products.

It is satisfying for several reasons. First, it is the marketing of the product, and not its merits in any way, that eventually sells it.[†] Therefore the product need not be particularly functional or aesthetic for it to be a commercial success. Once again, success in the field can be safely put down to the use of one's own particular talents to the full on behalf of the company; failure would, of course, be more appropriately blamed on the marketing division who have clearly not been doing their jobs properly. Second, the general standards of design are—though one would not wish to publicise this view to the Lay Masses, who would be bound to misinterpret the fact—somewhat ghastly (through no fault, we can be quite sure, of the profession). Therefore it is not difficult to produce 'breakthroughs' if necessary for reasons of personal gain or professional fulfilment. Third, it is very Fashionistic just now for college-trained (or *indoctrinated*, as the enemies of Nonscience like to call them) personnel to work in fields of Nonscience other than those for which they were trained. Thus one has pop musicians who trained as economists, writers who were trained to be mathematicians and designers who were trained at anything.[‡] So no matter what your own field of Expert conditioning may be, you are still likely to get on well in design. In the same manner that, in the United States, it is now a policy matter to have a black person somewhere on your staff, so it is becoming almost socially necessary to have an Expert in the British design team. This all works in your favour.

We may briefly consider a couple of examples of design by Experts, which testify to the fact that actually producing a *functional* article, in the old-fashioned sense of it performing a purpose, is no longer truly obligatory. The main purpose of design (and you should be grateful to the Author for recognising the fact) is to provide employment for Experts in the field. Second, design is intended to boost the financial solvency of the company sponsoring it, or whatever; third, it is a good outlet for modern techniques that otherwise might

---

* Except Experts.
† The student of the subject will recognise that this rule applied also to the successful perpetration of new discoveries under the guide of Nonscience. Here too it is not the merit of the discovery itself (Heaven forbid!) but the acumen with which it is assimilably communicated (i.e., *sold*) that matters.
‡ Anything except design, that should be.

never be used. These are such peerless motivations for the Expert that the out-of-date criterion of function—which is really a very basic and boring facet with which to bother ourselves—frankly ceases to matter anymore.

For example, to take a very basic example of modern research, we may consider the electric cell (or battery, as it is more popularly known, even though it isn't). Clearly a battery that is bright, modern-looking and definitely trendy has the edge over one that isn't. If it is faced with glossy enamel paint finishes, so much the better.

But if it is *leakproof* ...

Almost all the public, bless their hearts, little as they know about Nonscience, *are* aware that batteries tend to leak as soon as they go 'flat'. It is very hard to make a leakproof model. And this is where good design comes in.

Why would uneducated, uninitiated Lay Persons buy a certain type of battery? Because it didn't leak? No—after all, they have no means of knowing whether it does or not. They buy it because it SAYS it doesn't leak *on the label*. This is what makes the object commercially successful. Thus, from first principles, there is no need at all actually to make the thing leakproof, so long as it says it is.

In all probability the battery will leak eventually. When this happens you could be in trouble. So as a safeguard (and this is where design Experts come into their own) the casing is made of corrodible tinplate which will quickly decompose and so obliterate the wording about non-leakage as soon as the thing begins to leak. Thousands of people every year are caught by this one.

Another good example of how a simple design can aid the manufacturer is the water tap shown in the diagram. Old-fashioned taps were directed downwards, in the direction in which the water was intended to flow. But a tap with a leaky washer drips, so making an annoying sound. So the Expert steps in,

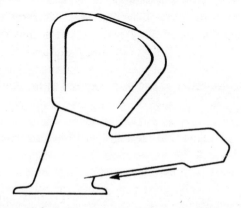

The ultra-modern Expert-designed faucet or tap. Note the carefully designed anti-drip upward turn of the nozzle attachment. For the significance of the arrow and its implications for Experts in other disciplines, see text.

and in one brilliantly successful move produces a tap which is ideally suited to modern housing in that it prevents this from happening. Altogether, and in one fell swoop, the problem is overcome—by the dramatically simple expedient of bending the tap upwards, as shown.

Now, if it drips, water runs down the stem and into the surrounding area. No noise, no annoyance, and no problem. Not only this, but the leaking water provides a ready source of corrosion, rot, seepage and other forms of deterioration to the surrounding area that in turn provide excellent means of employment for Experts in other fields, which is economically sound anyway. And believe me, it works a treat.

Earlier we were looking at the use of metric standards in designing, with all the extra profit that that involves for someone, and the Expert designer has limitless resources available if they want to alter our everyday standards in other ways.

The colour coding of electrical wires has already been changed. The three colours in England used to be RED (for danger, indicating 'live'); BLACK (for neutrality, indicating 'neutral') and GREEN (like the grassy ground, indicating 'earth'). Such pitifully basic colours were bad in two ways:

(a) they are clearly self-evident and easily remembered by Lay People; and
(b) they are very simple to manufacture.

So to bring us in line with a more Nonscience-viable future, we have scrapped those. Now we use blue for neutral, brown for live and—wait for it!—*green and yellow stripes* for earth. Lay People would never guess which was which, or if they did find out in some way they'd certainly never remember them correctly. It all helps to make the Expert feel wanted. Now think of other examples for the future. Green for danger, perhaps?

The still Fashionistic pollution field has two very propitious aspects to it which concern the Expert:

(a) we may do it;
(b) we may also find topical and exciting ways of claiming that we do not do it anymore.

The Expert is quick to realise, of course, that the word 'pollution' has many emotive implications. Yet the side-products of an expanding Nonscience-orientated world are inevitably going to accumulate somewhere. It was in 1970 that 12,540 nerve gas rockets, each containing enough material to gas some tens of thousands of people, were dumped into the sea near the Bahamas. And what an outcry followed! Yet what could have been more sensible? The gas itself was produced by the Nazis during the Second World War, and now we have far better and more efficient means of gassing civilians if the need arises. The old-fashioned gas simply *had* to be dumped, didn't it? And its disposal in the sea gives us much useful information on:

i. the rates of dispersal and hydrolysis of the gas if it reacts safely with sea-water; or alternatively
ii. the nature of the widespread damage if it doesn't.

One day (when we all live under the sea in air-tight chambers as a matter of course) it is perfectly possible that such a submarine attack might be launched by some unscrupulous enemy. The 1970 experiment will give us valuable information on the behaviour of these materials—data we could not have obtained otherwise.

Without any doubt a sure-fire winner for pollution research is going to be the heavy metals. Mercury and lead between them have been put forward as toxic elements leading to a whole range of conditions, and one could easily add to the list conditions such as:

- mental illness;
- baldness;
- impotence/sterility;
- cancer (of course);
- genetic deformities, etc.

Levels of methyl mercury in tinned tuna fish on sale at present go up to near the one part per million level, which is getting towards the level that would call for a ban on their sale. In the United States, Vermont and Michigan have simply banned all commercial fishing, because of the levels of contamination, and it has been suggested that Americans do not eat fish more than once a week, unless they are sure it came from deep waters. That is no real answer, of course, since lead has been found wherever the samples are taken in the sea.

The highest levels of pollution are found around the coast of Japan, and the metals are concentrated in crustaceans; then, as the next link in the food chain, they are concentrated still more in fish. Salmon and tuna are the main sources of heavy metal contamination in diet. But Experts could easily launch surveys of pollution levels and look in the literature for all kinds of diseases that it might be causing.

For instance, DDT (now the world's most Fashionistic target in the insecticide field) acts as a neuromuscular poison in insects, doesn't it? So in all probability it could be linked with mental disease, premenstrual tension, Parkinsonism and so forth. Almost any worldwide pollutant could be claimed, by an Expert, to cause virtually any condition in humans.

Publicity for the results would be immediate.

On a more global scale, studying the weather has excellent prospects for successful employment, for two main reasons:

(a) all the might of modern Nonscience can be brought to bear on the subject;
(b) but nothing at all practical can be done about it.

Even satellites have been orbited to keep an eye of the world's weather and supply photographs of cloud cover (and that's all, in spite of what the public is happy to imagine) to countries beneath. Britain has full access to these arrangements although, true to Nonscience at its best, forecasts were not based on these pictures when they became available but were worked out on the basis of past weather patterns. Examples such as the record rainfall in June 1971 (forecast as a record-breaking sunny month) show how healthily randomisated it all is.

An excellent example of just what an Expert *can* do in this field is the East Pakistan Cyclone towards the end of 1970. The entire course of this fascinating disturbance was monitored with aplomb and efficiency. From its birth in the Pacific, to its pathway across the Bay of Bengal, the progress of the cyclonic depression was carefully monitored by the satellites ESSA 8 and ITOS 1. The severity of the storm and the time at which it was expected to make landfall was predicted over 30 hours beforehand, and the Pakistani Experts—through their radar and wireless stations and satellite tracking units—kept a careful eye on its progress, an hour-by-hour survey of its gradual build-up to full force.

By the time the cyclone actually struck the Ganges delta itself its evolution had been subjected to a detailed study which, in the final stages, centred on the Pakistani radar station at Cox's Bazar. But what would it all actually mean in practice to the area? The answer lay in a report published in 1967 by the Red Cross, which stated that under these circumstances 1,000,000 people in the region would be casualties in the event of such a cyclone striking. In fact, the predictions were very good. The time and severity of the cyclone were right, and so was the unexpectedly high tide; it was all as predicted. Unfortunately the number of people who died was a little lower than the Expert prediction, which was the only noticeable discrepancy. But otherwise the whole exercise was a triumph for predictions—a wonderful example of Nonscience to the world.

Of course, if times get hard for you, an extra income can always be made by spare-time activities. I am not thinking of cleaning cars, serving in bars or any-thing like that,* but more academic and Nonscience-orientated occupations.

---

* As we have seen (page 91), this form of activity for graduates is not uncommon, but generally supervenes when they drop out of the Nonscience Movement altogether. A just fate for their negligence, one might say.

Once your students are fully trained and have *passed* the examinations, the most complex and involved problems can come their way.

Coaching for examinations is one such activity. Charge around £1 (£15 in 2020) an hour, and impress on the student's parents that lots and lots of hours are necessary. Use esoteric expressions to keep the pupil's respect, and enough humour to maintain his interest. Above all, convince them that they have to do considerable amounts of private study between this week's session and the next one. This will subconsciously make them learn what is required of them and this, along with the few elementary tricks of technique that you may impart, will guarantee their success. Always remember the cardinal rule: ANY IDIOT CAN PASS EXAMS. Even them.

Marking examination papers is something else with which to whittle away the winter evenings. This pastime (often known as being 'on the racket' in the trade) can bring in around 25p (around £4 in 2020) per script. One GCE examiner I came across used to make over £100 (2020 value £1,500) holiday money every summer by marking a few hundred papers in the fortnight before the end of his university session. If you can be appointed a Chief Examiner—and that relies on the Old Boy Network exclusively, I fear—you could make £500 (£7,500 in 2020) for the job of overseeing the other markers' work. That is an arbitrary and rather meaningless task anyway but it does add to the aura of officialness that should surround any competent examination.

The examination results are fairly precarious, and it is essential to avoid any comeback on them.

One way of doing this, if you think a reassessment is possible, is to destroy all papers after you have marked them. No questions can then be asked, and you can always state that:

(a) if you haven't been caught before, you did not know it was not allowed; or
(b) if you are an old hand, they were inadvertently destroyed/stolen/thrown out/used to light the fire/sent down to the fish and chip shop by Nanny or whatever.

The last time there was a thorough going-over of examination results was in 1936, when Sir Philip Hartog and E. C. Rhodes took 15 papers that had all been given exactly the same mark, and sent them off to 11 different examiners for re-assessment. Of course, they were asking for trouble. Some of the papers on the second marking were given percentages as high as 71, and others dropped to the 21 percent level. Worse still, in at least two of the cases, it was reported that papers placed at the *top* of one examiner's list came at the *bottom* of another's. Well, with the revelations of the Nonscience Movement now available, we can all see why it was.

From the student's point of view the marking of papers has an air of authority and calculated accuracy about it. This is necessary, of course, if they are to do the Movement credit. There have been other surveys in the past that have shown how an examiner will give the same paper different marks if he (a) marks it again later or (b) marks it at the beginning of a run, instead of at the end. But this is all part and parcel of our randomisation system. Another equally vital aspect of it is the tendency for different examining authorities to apply different criteria. There was one example in which the same examinees sat examinations under two different authorities (though in the same subject). There were 27 successes in one set of results, compared with only 3 in the other. The single failure under authority 'A' had grown to 25 under authority 'B'.

That is another excellent illustration. A further example concerned the batch of Worcester examinees who failed their examination. When an enquiry was made it was found they had all *passed*, instead.

But now that you are on the donating, rather than the receiving, end, let's go into some detail about examinations. The essential principle is that of Categorisation. Not the ordinary pigeon-holing of people into convenient groups, but the drawing of a hard-and-fast line between the haves and have-nots. Let me clarify this. If we select any population of objects, we inevitably tend to group them in our minds. Thus if we talk of height, we may classify people as being short, average or tall. In the same way we tend to think

PATIENTS do not take in much of what they are told by their doctors. In a survey of outpatient communication carried out at the London Hospital, patients remembered less than half the information they were given.

However, paradoxically, some managed to "remember" instructions exactly opposite to those in fact given them.

The British will build faster tower blocks if they measure them in millimetres, trees will grow taller, cows will produce more milk and better beef, tomatoes won't be soggy if they're metric, hens will lay bigger (and perhaps fresher) eggs. It'll be faster to drive in kilometres, there'll be fewer accidents on metric roads, and less drunkenness if people down their Guinness in half-litres and not in pints.

NOW A TANGELO . . .

Rome, Monday. — Citrus growers in Lentini, East Sicily, have developed a tangelo — a high vitamin cross between the tangerine and grapefruit. — Reuter.

LONG-NOSED BEE

Soviet scientists have developed a new variety of bee with a longer probiscis which increases its honey-gathering qualities and its resistance to cold. Moscow Radio reported. — Reuter.

Kinesics is a new science—the study of body language. Its students want to know what we tell each other with a nod and a wink and what we mean by crossing our legs.

Often we don't realise we are doing such things and occasionally our body language contradicts our words.

UNIVERSITY lecturers are on to a good thing—they teach students for only a few hours each week, and their holidays last for half the year.

The current experiment at the applied psychology unit is using 16 TV screens. On each screen a different part of a prison is presented. With the advice of a Home Office representative, realistic "incidents" have been filmed. One of these, for example, shows a man stopping outside a prison wall, measuring the size of the bricks, making a note and then walking off.

The attitude of the patient towards the physical examination may be revealing. For example, hesitation prior to insertion of a vaginal speculum is usual; vaginismus suggests an up-tight attitude.

SEXUAL relations between father and daughter may be "accepted" behaviour in certain social sub-groups and need not imply sexual deviation, paedophilia, on the part of the father.

This suggestion was made by Dr. Narcyz Lukianowicz, consultant psychiatrist at Holywell Hospital, Antrim, when he described twenty-six cases he had come across in Northern Ireland

A private practice physiotherapist described his "spectacular" and "almost miraculous" results in the treatment of such conditions as a black eye, arthritis, a split inter-phalangeal web, varicose ulcers and 15 cases of ankylosing spondylitis.

Square pineapple

A square pineapple has been developed in New South Wales.

A CALL for a study into the apparent link between academic ability and myopia in children has been made by Nottingham's Principal School Medical Officer, Mr. F. E. James.

DELIVERING a lecture once, the psychiatrist Dr Ronald Laing made his audience sit up with the opening line, ' The initial act of brutality against the average child is the mother's first kiss. . . .'

A test was done on a group of Cambridge housewives, in which a series of tape-recorded words was played to them through a background screen of noise. The word most frequently heard was " peace " with the word " bed " a close second.

The single women and most of the single men brought partners. For those who did not, the foundation provided surrogate partners.

Depending on the type of dysfunction being treated, the physical techniques are gradually advanced until orgasm can be successfully and repeatedly achieved.

A 17-YEAR-OLD typist, worried about her mother's illness, was said yesterday by the Shrewsbury deputy coroner, Dr. D. A. Urquhart, to have died of a rare condition, difficult to diagnose by examination even after death.

"It is a combination of mental and physical illness whereby the bodily resources are worn away to an extent where they are not compatible with life," he said.

Students may be guided in their choice of a career by these interesting extracts from the Lay Press.

of people as being thin, normal or fat. You have a car that is a big car, or a mini or a family car; and a spouse who is miserable, all right or superb. Thus we group things, instinctively, and we bear in mind always this concept of normalcy and exceptions from it. The same applies when we talk of ability.

Instinctively we group people, when we meet them, as intelligent, ordinary or dull. And these innate tendencies of ours are basically quite realistic.

But this kind of altruistic realism defeats the aims of Nonscience. We do not want a system that says someone is a bad Expert, an average Expert or a brilliant Expert. What we require (and what our examination system grants us) is the ability to designate them, clearly and unequivocally, as *an Expert or not*. Simply and concisely, just like that.

So we have the Pass Mark. This is a level* at which the line is drawn. Those candidates who are given marks above the predetermined limit *pass*. Those who come below this line *fail*. Thus we have a clear-cut distinction. The results are best seen if we tabulate an example.

Let us consider four hypothetical classmates: Bert, Alan, George and Laurence. Friends would say that Bert was frankly uninspiring and really fairly thick. Alan and George, it was said, were average students, alike in most respects; while Laurence was a run-away genius, who was bound to do well. The exam results come along. Bert clocked up a miserable 14%, a useless but predictable result. Laurence gained a well-earned 94%, the highest mark of his year. Alan and George were separated by only a single mark, much as you might expect. Alan was given 50%, George 49%. And so we find that the results in this hypothetical example go along with what one would have expected.

But this is where Nonscience steps in. For by drawing the line across at the 50% pass mark, we have a different situation altogether. According to this, Laurence and Alan are together fully qualified Experts; while George and Bert are documented failures. Thus we have a complete re-grouping, from the three categories that we would ordinarily utilise into the black-and-white realities of Applied Expertism. According to this system, George is no more qualified than the man who sweeps your yard or the woman selling bootlaces.† But Laurence and Alan alike are both certified as competent to run our lives, dictate the progress of civilisation, arbitrate over such interesting peripheral issues as life and death of the individual and control the global environment. The fact that George and Alan were separated by only a single mark is neither here nor there when we get down to it and, so implicit is our faith in examination results, that no-one would dare to question them in the ordinary way.

Of course, we must bear in mind that in different years the average ability of the students would vary anyway. And if a given percentage pass, then the status

---

* The Pass Mark is sometimes 50%, sometimes 45%, sometimes 40%, occasionally 60%. But this doesn't matter at all, since the examiners always correct the sample so that the desired percentage pass. If you present a paper which the examiner decides ought to just scrape through, then you would have 51, 46, 41 or 62% respectively, no matter what the absolute standards might be.

† It is possible that these people may be highly qualified Experts who have been unable to find employment, of course (page 91). So do bear it in mind.

that George and Alan will gain must inevitably alter from year to year, and from college to college. Had they been in a slightly lower standard of classmates one year then both would have been qualified Experts after the examination: on the other hand, if the average standard had been higher, both would have failed, and left with no more claim to academic status than the person who had never sat the examination or been to college in the first place.

Study of the student's abilities and unravelling the mental processes involved brings us into another exciting field of Nonscience—the study of the mind.

There are two important disciplines in which Experts are concerned with the human mind. The first is psychiatry, which need not detain us for long. Psychiatry is a form of mental disorder in which the sufferer feels they have a profound insight into the mental problems of others. The patient—the *psychiatrist*—generally sublimates this tendency by the use of the trappings of Expertism and Nonscience. It is not a very healthy business, this, but at least it causes no great harm and it is certainly arguable that the clients of the psychiatrist feel some transient benefits from their presence, and from this apparent ability to understand their problems. The psychiatry Expert, on the other hand, obtains gratification from their contact (in an Expert-and-subject manner) with the client. In this way they both end up happy.

Psychology is different. Whereas psychiatry is based on abnormality, psychology is the invention of codes and models with which to integrate normal mental behavioural patterns. But it naturally has a lot to do with abnormality, too.

There have been many Experts in this field in the past, and since the subject has such intimate connections with the Lay Person it is easy to see why their names have become almost household expressions. Freud is an example. His work is a supreme example of Nonscience at its most refined.

Freud was an introspective young man who carried on unsatisfactory relationships with his wife's younger sister and had a most stultifying mother fixation. In realising—not for the first time!—that unconscious motivations play a profound part in decision-making and human behaviour, he assumed that perhaps the disorders he manifestly suffered from applied to everyone else around. And this approach has founded many branches of our discipline, so the aspiring Expert could bear it in mind. He could—depending on his own particular kinks—see everything in terms of latent homosexuality, fetishism, nail-biting, toe-clicking, constipation phobia or whatever. Latency, as it may be called, can itself expand into a new branch of Nonscience before long.

For example, it is clear that a young man from a happy home background will tend to choose as a wife someone who closely corresponds to his mother. If his mother has always been successful, and has lived up to the ideals he sets for his own future, it is patently obvious that he will instinctively try to find someone similar for his own wife. Now at this level of argument we are talking truisms; there is not much Nonscience about it.

But wait. It is perfectly simple to construct a philosophical model around these basically observed facts. For example, let us suggest that this represents a sublimated sexual assault against mother. Or let us postulate that the young man, through this sublimation, really wishes to *marry his mother himself*! Now we have it—a far more incisive, interesting conceptual model than the mere 'truth' itself could ever provide.

His choice of a fiancée now becomes a frustrated attempt to make love to his mother—and we have the basis of a sound and thoroughly acceptable piece of Freudian Nonscience.

Again, we might observe the strength of the copulatory instinct in animals and deduce that this is instinctive, a necessary adjunct to survival. But let us apply Nonscience to the matter. We may now postulate that the joining process is a *desire for the male to return to the womb*. Is that not a far more entertaining thesis? The bare 'truth' itself can do little but inform. But this theoretical approach gives us a subject for discussion and debate, and a range of possible research projects too wide for the Ordinary Mortal to contemplate.

Clearly there is unlimited scope for future activity. Unofficially one can bear in mind that we have not the slightest hint, as yet, of how the brain functions; where are its main centres of activity; how it communicates between its different parts; what the different parts are or where; how stimuli (such as pain) are transmitted, received, interpreted, stored, recognised, reacted to, or what they are in the first place. There is no clue whatever to the means by which stimuli are transmitted into impulses, so almost any theory can stand up on its own in that field. Neither do we have any grasp of the functioning of a single neurone (or nerve cell)—but I will give one clue. In the past it has been fashionable to talk about 'go or no-go' in this field: i.e., either a neurone triggers or it doesn't. The analogy is with the computer and its 1 or 0 mode of operation. But though this simple concept suffices for many hypothetical models, it is very obvious that the neurone is more complex than this. When single-celled protozoan organisms can embody the functions of sight and smell, movement, reorientation, attack and defence and so on, it is clear that a human neurone can do more than merely fire an impulse or not.

But of course no-one has any knowledge about this in practice, and so for a good many years yet the field is open for suggestions. And since they could be said to *affect the very basis of the human mind*, no matter what the suggestions may be, it is clear that Experts in psychology can have a confidently secure and interesting future.

On the purely theoretical side there is plenty of scope just now for research and investigations into symbolism. Symbolism in the media, in design, in our lives; with particular reference to phallic symbolism, of course. Quite recently the Expert view has gone as far as to suggest that phallic symbolism has dictated our construction of umbrella handles, for instance, and factory chimneys. Sportscar streamlining is phallic symbolism too, and so of course is the spear. What this means in practice is not, of course, clear; but the purpose of Nonscience is to elaborate, and not to clarify.

But the further potentialities of phallic symbolism are considerable. For example:

- Consider symbolism in writing. The use of the capital 'I' instead of the lower-case 'i' to mean the assertive first-person singular is an interesting English convention. Clearly, the Expert might argue, the capital 'I' is *tumescent and erect*. The carved stone-age warrior on the hill at Cerne Abbas, Dorset shows him assertive, sexually erect and aggressive. Surely the capital 'I' is a literary symbol of exactly the same nature. Similarly, does not the letter 'B' depict a scrotum containing two pendant testicles? And is it a mere coincidence that words such as Balls, Bollocks and so forth contain the device? Certainly there is a research project here.

- Extend the evidence for phallic symbolism in the world at large. If factory chimneys (which the mere Lay Person assumes to be functional in design) and sportscars (which the uninitiated believe to be 'streamlined', of all things) are phallic symbols, then so also are a great many features of our modern world. Why, the insertion of a program into a computer is clearly symbolic intercourse, not to mention the jerky, thrustful electronic jiggery-pokery that goes on afterwards and the orgasmic spewing out of the final printout tapes. Picking one's nose is symbolic defloration, in public it becomes sexual exhibitionism of a peculiarly private nature; toad-in-the-hole (the very name rings of sexual symbolism) becomes a dish of the most sensuous kind; and oh! the hidden meaning behind pulling the knob on the lavatory cistern and the torrential rush of flushing fluid that results.

One might even postulate that man himself is a phallic symbol, this being why polo-necked sweaters are so popular. The foreskinlike rolled neckband is the giveaway here. In this case, by simple extrapolation, we may conclude that the practice of cunnilingus is no more than an attempt at intercourse with the largest symbolic phallus available for that purpose.

Sex itself is a highly Fashionistic field of Nonscience endeavour. Some of the headline-catching pioneer work (the first Expert analyses of the sex act, etc.) have already been done. But there is plenty more in train.

For instance, it has become Fashionistic to divorce sex from its emotive context and treat it as a physical act, pure and simple. (Indeed, the intensely Fashionistic aura with which the subject has become surrounded means that, for instance, marriage is seen as a dying phenomenon. Try not to disclose if you are married. Or if you *are* caught out, indulge in wife-swapping. No matter if you both dislike it, or even if it costs the marriage; better that than a Fashionism Failure, after all.)

Sex education is a popular refinement. It is now held that children need not merely be liberated from shyness about sex, but have to be trained in the whole thing even if they lack the physical wherewithal to do it. The most exciting and elegant example in recent times was the film, released in the summer of 1971, for British schoolchildren, showing among other things naked men and women masturbating, and intercourse taking place.

That was a masterstroke. It was obviously going to make history and catch headlines. And now that the trend has been established, let us see what else we could suggest.

The important basic properties of the film are:

(a) it shows phenomena that occur naturally in humans, whether previously shown in films or not;
(b) the acts themselves are seen as a 'natural release' for many people and
(c) they are intended for a school audience which is clearly coming to the subject in its filmic context in order to be taught.

Now, there are many other ways in which this splendid precedent could be followed. Babies could usefully be shown films of normal toilet behaviour, ready for when they are old enough to leave their nappies and diapers behind. The film should show the normal passing of urine and faeces, with specially illuminated close-ups of paper-wiping procedures. Correct ways of sliding underwear garments *up* and *down* should be highlighted, ready for when the youngsters start to do it themselves.

Higher up the age scale we could make arrangements for vast teams of Experts to prepare films on other ways in which people find 'release', from drug injection methodology and fetishism to homosexual techniques for the beginner and Applied Fidgeting. Other courses could cover the biting of nails, the picking of the nose (with methods of nibbling the nostrillar contents as an optional extra) and how to squeeze pimples. Courses in methods of scratching the anus effectively could be added, along with detailed sequences showing the normal range of actions used in fiddling with one's genitals in an idle moment.

The mind boggles at the potentialities. Why wait? Strike now whilst Fashionism is high—the world's headlines await you! It's all part and parcel of progress, remember; and with the cloak of 'psychology' it's clearly the way that all good Experts must follow.

As these examples show, there is limitless room for new development. Sexual symbolism is only one part of it, of course. The meanings attached to other characteristics of the environment are equally available for examination and original comment. Shapes, colours, preferences for texture; the combinations are unlimited.

This leads us to the last, and perhaps most propitious, avenue for the psychological Expert—the manufacture of what are called psychological tests.

There have been many in the past, of course. This has certainly led to a need for greater cunning in the future, but not to worry. Ink blots, fuzzy pictures that seem to show a face, simple outlines, key words, are amongst the forms of test that have been used. But with number- and letter-selection (see paragraph on sexual symbolism in the alphabet, page 243), picture- and image-sorting, multiple choice of diagrams and outlines, preference for fruit and vegetables of different types, liking for names and so forth there is no shortage of research material in the future.

The most important thing to realise is that Psychological Nonscience is quite popular just now, and is even enjoying something of a vogue. As in the other branches of the Movement, there is no need for repeatability or an objective means of assessment of the results. The Expert who thought up the test is responsible for the marking procedures too, and Experts are simply not questioned over things like that. So if this is your bent, press on.

Do not be discouraged by meaningless fears and pointless introspection if you are approached by industry to act in senior staff selection. The people you have to choose are destined to become Experts in their own field, too; and yet they have not reached their position by the endless years of training and processing that you have had to endure. So it is only fitting that they should have some means of selection before going any further.

And, in choosing the mature Management Expert, what better way of ensuring their suitability than by calling in an Expert to select them? So press on boldly. If they do well, it will be seen as a vindication of your balloon test, sausage-bending procedure, spider-drawing criterion or whatever you decide to utilise instead.

If they fail, put it down to personal deficiencies on their part. The exception, remember, proves the rule.

In that field too, you are working *close to the Ordinary Lay Public*, and this is important. Experts always prosper better, faster and more successfully if they keep their field of concern somehow close to the Man and Woman in the Street. You may notice that reporters keep asking you:

'What difference will your discovery make to the Ordinary Member of the Public?'

and if you are foolish enough to answer:

'None whatever'

you can be quite sure that the story will never be used.

It is imperative, therefore, to *think of a link*. You remember the electron microscope that could 'see atoms'? Such a very clinical, detached, impersonal development that was (page 166); but with the claim that it might be used in the discovery of the *final breakthrough against human cancer* (see Key Phrases, page 169) it became International News of the Highest Order. Do not state that you are merely investigating some foodstuffs, but word it to emphasise that your research may make a difference to the way that Western men, women and children eat their food. It *must* relate to the Ordinary People.

The most obvious and simple means of doing this is to join that branch of Nonscience known generically as sociology. It is concerned with the observation and documentation of the day-to-day activities of human individuals, but written and described in the terms of the movement. Thus you should describe yourself, not just as a 'sociologist' or (better) an Expert; but as a 'Socio-economic Investigationary Activist' or a 'Human-relationship Analytical Documentationaliser', to give just two possible examples. There are scores of possible fields that you can enter and the prospects for the future are excellent. In fact, as long as there is Nonscience to alter our world, there will be sociologists to study the effects of it all on humankind (for an example see spina bifida, page 82).

For examples of the simplicity of this subject, let us turn the pages of a recent Seminar on Poverty organised in the United States. This account provides some quite typical standard approaches to the definition of Lay Terms such as:

A slum: 'the allocation of a very limited portion of a person's resources, abilities and energies to the ownership, maintenance and adornment of residential structures.'

Dad is away: 'a set of arrangements for producing and rearing children the viability of which is not predicated on the consistent presence in the household of an adult male acting in the role of husband and father.'

This is pretty mundane for many Experts, of course, and they prefer to deal with the Lay Public at a level where they not only observe what is going on, but actually take steps to provoke widespread changes in society. That is the power of Nonscience at its most gratifying.

Members of the public are notoriously susceptible to findings that affect their stomachs. Food and diet in general are always Fashionistic. And there are many ways in which the crafty Expert can turn an honest penny from this particular speciality.

For instance, cyclamates have been banned as sweeteners recently on the basis of some experiments in which they were given to rats (along with certain other ingredients) in extraordinarily large doses. The rats died. Public outcry over the results was international, and the fame that resulted for the Experts involved was considerable. At the centre of the argument was the suggestion that: 'These materials have been released for public consumption without a full investigation into all their possible side-effects.' This is a principle that can be extended into other fields.

(1) Among foods that are known to cause human disease if ingested in sufficient quantities are:

WATER   KIPPERS
VITAMIN D   TEA
FATS   COFFEE
ONIONS   ALCOHOLIC DRINKS OF ALL KINDS
POTATOES

(2) Some exciting results could be gleaned by studies of the following food materials, all of which can cause instantaneous death if injected intravenously into the experimental subject:

BREAD   SUGAR
BUTTER   MINCE PIES
FRUIT CAKE   LIVER PÂTÉ
PORRIDGE   CAVIAR
CORN FLAKES

(3) Many foods have been released for public sale *without a detailed assessment being made of their possible side-effects* such as:

TEA   FRIED EGGS
SANDWICHES   OLDE ENGLISH MUFFINS
SAUSAGES   LEMONADE
BEEF STEAK   TOASTED TEACAKES
TOMATO PURÉE

(4) And there are many recent additions to our diet which could be described to the Lay Public as *'foods that have never been investigated for possible long-term risks'*, for example:

HAMBURGERS   INSTANT MILK
HOT DOGS   FREEZE-DRIED PEAS
FISH FINGERS   BAKED BEANS
CREAM OF MUSHROOM SOUP
VACUUM-PACKED TURKEY ROLL
DEEP-FROZEN SCAMPI

Almost any foodstuffs could be 'condemned' in this fashion. The publicity ratings for such a scheme are excellent, and so are the possibilities of perks from vested-interest parties.

STORAGE GUIDE
BY BEE NILSON

How long should you keep your food? In each row the top line indicates safe storage time in your larder, line 2 your refrigerator, line 3 your deep freeze.

1 **Cool larder 12°C.**

2 **Refrigerator 2° to 8°C.**

3 **Freezer −18°C.**

SMOKED FISH

foil-wrapped
1 day

2 days

6–12 months

RAW FISH

under meat cover: from shop use same day; fresh caught 1-2 days

wrapped foil or polythene 1-2 days longer if fresh caught

3–6 months

JOINTS

2 days

4–5 days

ham 3–4 months, pork, lamb, veal 6–8 months, beef 12 months

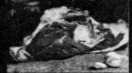

STEAKS AND CHOPS

1 day

3–4 days

as joints

STEWING MEAT

1 day

2–3 days

as joints

COOKED JOINTS

foil-wrapped or under meat cover
2-3 days

in foil or polythene
3-5 days

1–2 months

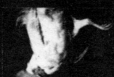

WHOLE CHICKEN

fresh: 2 days, thawed or chilled: not recommended

fresh: 3-4 days: thawed: 2 days chilled: unwrapped 3-4 days

fresh: 10–12 months

TURKEY

3–4 days

1 week

10–12 months

GREEN VEGETABLES

24 hours

washed and in polythene bag or box 5–7 days

6–9 months

PEAS AND BROAD BEANS

unshelled 24 hours

shelled and in box or bag 24 hours

6–9 months

SALAD VEGETABLES

in covered container 24 hours 1

washed and in airtight container 3–7 days 2

not recommended 3

THE SUNDAY TIMES MAGAZINE KITCHEN COMPANION 15

← KEEP THE PUBLIC GUESSING BY NONSCIENCE

- A splendid example of obscurantific explanationism from *The Sunday Times*. Note how the quoting of precise temperatures gives a sophisticated impression right from the start. How many Lay People know what they mean, or have the slightest idea of the temperatures in their own kitchens? You are right: none. Ordinary people, who imagined that fish were smoked in order to preserve them, will be mystified by the observation that fresh fish apparently keeps for longer, *unless it's in a freezer*. Then the roles are reversed. And they will wonder why kippers, having been smoked a long time before, will suddenly be in need of foil-wrapping—and even then will last for only a day.
- The use of absolute figures helps to confuse again with meat. Why does a joint keep for twice as long as meat for stewing? Indeed, how does the meat know what's going to happen to it, so that it may alter its behaviour accordingly? Lay People will naturally wonder why cooked meat (which is far more easily decomposed than fresh) keeps for longer. Their respect for the Expert who understands these things is thereby deepened.
- Note (and this is important) that a turkey will keep for twice as long as a chicken. Lay People readily realise that it is beyond their meagre mental resources to know why.
- And the final crunch comes with vegetables. Ordinary folk will be assuming that peas and beans keep for weeks—many of them will have done just that in the past. But they won't, says the Expert; they will last (like the legendary butterfly) for one delicate day. And putting them into the refrigerator doesn't increase that by even half-an-hour. The other date discrepancies will disturb them too: thus, why does a cabbage last for as long as a pea on the shelf, but seven times longer in a refrigerator, particularly when the Lay Person's experience may tend to indicate the exact opposite?

Actually, almost any aspect of our daily lives can be a candidate for Expert exposure as a 'hazard'. Lying in bed can induce hypostatic pneumonia and fluid stagnation in the lung. People die of the effort involved in love-making (which is itself a disease-spreading phenomenon) and sex is already leading to global overpopulation, starvation, chaos and misery—unless we use the Pill, of course, but that can always be incriminated on grounds of its own (page 213). Once we are out of bed there is danger in a cup of tea or coffee. Both contain dangerous alkaloids and tea has measurable traces of the carcinogen (i.e., a cancer-causing agent) 3:4 benzpyrene. Coffee itself can produce pancreatic stress which can cause, or exacerbate, diabetes.

At breakfast we find chances of typhoid from eggs, there is 3:4 benzpyrene again in kippers, the artificial ingredients in bread are often condemned for various entertaining reasons, butter may encourage a coronary heart attack … monosodium glutamate can cause convulsions and collapse and even salt is composed of the elements sodium and chlorine (one a dangerously explosive metal; the other a lethal First World War poison gas).

The emotional stresses of love and marriage can induce psychoses in some instances, even total mental breakdown. And the working environment is either materially hazardous (to those at the lower end of the social escalator) or—if you advance to the dizzy heights of executivedom—leads to stress syndromes and illnesses including strokes, heart attacks and serious stomach ulcers.

Certain highly trained Experts have been entrusted with the sophisticated business of research in dietetics and food preservation. One can but hope that the Lay Masses appreciate all this effort.

Indeed, the application of the ideals of Nonscience can go much further than this. Air's main constituents are nitrogen, oxygen, carbon dioxide:

- NITROGEN can induce narcosis and has been known to kill;
- OXYGEN leads to anaemia if breathed for long periods, and can also cause pneumonia in experimental animals; and
- CARBON DIOXIDE is a suffocating lethal vapour.

So even breathing could be described as a highly dangerous operation.

Even our water is no longer pure, we are told—and if it was there is still the well-documented hazard of cerebral oedema if you drink too much. And that can kill, too.

So for True Experts, there is limitless scope. They could, with the authority due to people of their standing, justifiably condemn eating, drinking, thinking, sleeping, sex and even breathing for what are plainly documented reasons.

The way is clear. Already many similar moves have been successfully accomplished.

May the future be bright and rewarding—Experts of the world, your time is *now*.

FIND THE RIGHT FIELD

Looking back to 1970 from 2020, we can see that choosing an area in which to ply your trade has changed a lot in the past half century. The 1971 book was absolutely right to say that plastics research was passé. The plastics developed in the 50 years before the book included PVC, polythene, nylon, polyurethane, neoprene, Styrofoam … about 20 altogether. Since the prediction? None. We do now have liquid crystals (but they aren't plastic). Carbon fibre, which was fading out when the book appeared, was swept up by popular industry and, rather than engines, they now make entire planes out of it. Passengers don't realise it, but modern airliners are made of plastic, like a soap dish. So the development of new types of plastic has ground to a halt; now campaigning *against* plastic is the way to go.

Precisely as predicted, marijuana did become progressively decriminalised, and it has given rise to a range of industries. One of the constituents, cannabidiol (CBD) has been incorporated into hundreds of products. As this book goes to press, there are official investigations under way because most of the products don't contain any CBD at all (or, if they do, it's only the merest trace). That isn't important for us; what really matters is that there's lots of money to be made. Prostaglandins were also predicted as a money-spinner, and in the 1990s they suddenly erupted into Fashionism. They're everywhere now. Cash in!

FASHIONS CHANGE

In 1971 the contraceptive pill was being linked to a range of problems. It was new then, and topical. Now, the same kind of side-effects are being linked to statins instead. This is the point: whenever something new comes along, the same range of everyday problems are assumed to be 'caused' by it, whatever it is. This theme runs throughout Nonscience and is always a sure-fire way of making headlines. If things do go wrong, you are guaranteed immunity from prosecution—as we saw earlier, so long as you state publicly: 'lessons have been learnt' then you're going to be fine. The section in the earlier book on forensic science was particularly prescient. Back in the day, in the era of Sherlock Holmes, people knew that forensic investigation would find out the truth. Now it is different—the point of a forensic Expert is to secure a conviction, not find out what happened.

One particularly intriguing item in the book was the table of conditions seeking a cause. Even after 50 years, there's only one change to be made—take out 'spina bifida' and insert 'dementia'. In spite of what people say, spina bifida has not been completely resolved. Folic acid in the diet plays an important role, and 400 micrograms (mcg) of folic acid should be taken every day throughout pregnancy. But it isn't a complete answer. It reduces the incidence of neural tube defects by 70%, but it does not reduce them to zero. This is odd. Vitamin C gets rid of every case of scurvy, and Vitamin B12 will cure every case of pernicious anaemia, but Vitamin B9 does not eliminate every case of spina bifida. Hmm. Substituting dementia in its place keeps our table current. And note that the list already had Alzheimer's listed (at the time, it wasn't known to the public).

There was a section about the recycling of waste, and that has become big news. In particular, everybody is against plastic. Between ourselves, this is profoundly unbalanced (as are most aspects of Nonscience). Plastic has given us benefits that were previously unimaginable. Producing the alternatives (like paper, glass, and steel) liberates much larger amounts of greenhouse gases than plastic. The products we can make from plastic have crucial roles to play in every aspect of our lives and, without plastic, human societies could not function. Toys and storage facilities, medical apparatus and everything electrical … our homes, factories, offices, schools and hospitals would all fail without plastic. The reason for the condemnation is—plastic in the oceans. More than half of that is dumped by the very people who rely on the oceans to survive (namely, fisherfolk; I was going to say 'fishermen' because I have never seen a female crew member of a fishing boat, but fisherfolk is the term we need) so the real people to blame are those who are the actual custodians of the sea.

Campaigning to save the planet reveals strangely mixed motives. Lewis Hamilton is campaigning for us all to protect the environment; he is a record-setting racing driver who makes his money by driving gas-guzzling racing cars. Harry, Duke of Sussex, who was at the time a member of the British royal family, gave an impassioned address at a climate change conference with his feet bare, to show how genuine he was, travelling by private jet and arriving by helicopter, with massive super-yachts and over 100 aircraft ferrying in other delegates. Meghan, who later published her delightfully petulant autobiography *Finding Freedom*, resents any criticism of her eco-friendly lifestyle; whether it's flying to her $430,000 baby shower by private jet, or being driven around Los Angeles in a Cadillac Escalade SUV with its six-litre engine. Unsurprisingly, she was uncomfortable during her brief excursion into the civilised world of royalty, though it did mean she could swap her modest 3-bed home when she filmed for cable TV in Canada for a colossal mansion in Santa Barbara with nine bedrooms, sixteen bathrooms, a games room, gym, and separate tea house. It doesn't seem to have any solar panels. I expect they'll come later. The actor Benedict Cumberbatch took part in demonstrations in Trafalgar Square, London, for the Extinction Rebellion movement while featuring in advertisements for the highly polluting MG GS sports car that he drove around Trafalgar Square. In a demonstration in the summer of 2019, several members of Extinction Rebellion glued themselves to trains in London to demonstrate against climate change. That's odd: those light railways are the most efficient of all modes of mechanised transport. The trains to which they were glued are the best answer to climate change. If the protesters wanted to make a

Distressing images of sea creatures trapped in plastic have provided much of the impetus for the anti-plastic demonstrations around the world. However, these animals are not trapped in waste discarded by the public, but by nets abandoned in the sea from fishing vessels. By far the greatest weight of plastic in the sea comes from this industry (it may be 70–80% of the total). Has anybody seen people protesting against that?

Much excitement was triggered in March 2020 when Experts announced the world's first tanker for carrying liquid hydrogen, a completely non-polluting fuel. Less was said about where it comes from. The energy to split water into hydrogen and oxygen at Latrobe Valley in Victoria, Australia, is created by burning colossal amounts of lignite (brown coal). This is the filthiest and most polluting fuel on earth, releasing five times as much CO_2 as the same amount of gas, and polluting the environment with toxic heavy metals and poisonous oxides of sulphur and nitrogen. The 'clean' hydrogen-burning vehicles in Japan are actually the most damaging to the environment anywhere in the world, so let's hope nobody realises the fact.

point, they should've stuck themselves to the prow of fishing boats heading out for a month on the stormy seas. I didn't see anybody doing that. These campaigners just complain. They never show what we should be doing. If they never rode in cars, and always hand-washed their clothes (in cold water) I might be more impressed. But complaining and demonstrating is all they want to do, not actually solving any problems. They offer no example to anyone.

Apart from fisherfolk dumping plastic in the sea, most of the rest is dumped by the public. The way to curtail the problem is so easy: never let plastic escape into the environment. Job done. Whenever there is a massive demonstration against plastic, what is left behind? Mounds of plastic waste dumped by the protesters. There is a lesson in there somewhere.

Not a single scrap of plastic should *ever* be dumped. We used to crap in the streets, but we learnt not to do that, unless it's very late and all the clubs are closed. Once people emptied chamber pots out of windows, littered the dining room floor with half-eaten food and left dead pets in the gutter, but eventually we learnt not to. It's the same with plastic. Stop dumping it, people. It isn't just industry's fault. It's yours.

The latest area of research is searching for microbes that can biodegrade plastic. The ultimate aim is to find some kind of microscopic fungus that could digest any sort of plastic, anywhere it had been dumped. Whoever discovers the miracle microbe that degrades plastic will, without doubt, receive the Nobel Prize.*

---

* I am not sure which one. There are Nobel Prizes for chemistry, for physics, physiology or medicine, and peace. There isn't one for biology, even though this is the most important science of them all.

Few undeveloped nations share our sense of concern for the environment. They are more preoccupied by trying to stay alive and saving enough money to send their children to school, or to visit the doctor. Many of these countries have only one way to dispose of plastic waste. They dump it into the sea when nobody's looking. You never see Western protesters in these impoverished places.

In confidence (don't tell anybody I said so) this would herald the end of civilisation. It means that all the electrical insulation everywhere would be under threat. Every water pipe in the world, including those in our modern cities and the newly installed pipes bringing water to the starving people of the desert nations, would collapse. Food containers would crumble, bottles spring leaks, medical apparatus would collapse, computers short-circuit, cars stop in their tracks, planes break up in the sky ... and it would also add greatly to the burden of carbon dioxide in the air, as every plastic item broken down biologically would pump more greenhouse gases into the atmosphere. The anti-plastic protesters never think about that. In Germany, Jasmine and Benny Zelos featured in a television programme on life without plastic. Viewers saw them throwing away toys, tubes of cosmetics, toothpaste, and toothbrushes. 'The first thing we put in our mouths every morning is—plastic!' complained Jasmine. They decided to switch to bamboo toothbrushes. The problem is that the bamboo goods we buy are actually impregnated with phenol formaldehyde plastic and the brush bristles are nylon anyway. These brushes deteriorate and have to be thrown out. A plastic toothbrush lasts for years. As the parents threw out the children's favourite toys, their daughter Nina was desolate: 'In the bath, you can play with stainless steel bottles or spoons instead,' said her mother. Nina was not impressed. Using plastic toys (and passing them down the generations) is the most ecologically sensitive thing you can do. By throwing out the plastic they were simply adding to the burden of waste which the authorities have to dump.

Two-thirds of plastic cannot be recycled. But what we should do is re-purpose it. We could separate all plastic refuse and keep it separate. Then grind it up and use it as ballast in concrete and asphalt—it could be used to build homes, factories, motorways, playgrounds. In this way, the plastic waste would perform a useful function; it keeps its carbon content stored away, it adds no burden to the environment, it doesn't have to travel far, and its use reduces the amount

This picture went around the world, to show how much plastic the pollution protesters left behind after a climate demonstration in Sydney, Australia. What a shocking sight! Hang on, other sources say it was left after the 2019 Extinction Rebellion protests in Britain. (Both wrong. It was taken after the '420 Marijuana Festival' in London's Hyde Park on 28 April 2019, but don't let mere facts stand in your way).

of energy we waste in digging out limestone ballast from the ground. Plastic in landfill is an excellent raw material for the future; re-used as I propose it can't go away, it won't escape into the ocean, and it has plenty of potential applications that conserve raw materials.

Although we could recycle one-third of plastic waste, in the United States 96% is not recycled. The world-wide recycling average is just 10%. Using traditional objects rather than plastic alternatives may seem instinctively attractive but the energy cost and the burden of environmental pollution in making a mug out of china is 100 times greater than manufacturing its disposable counterpart. Yes, the ceramic mug lasts longer, but every wash consumes more energy and releases polluting detergents into the environment. People like the idea of reverting from plastic to steel cutlery, but steel production causes vast amounts of pollution, which increases each time an item is washed. Paper cups may seem preferable to plastic, but it isn't always so simple when we calculate the environmental burden they impose.

The plastic items we use save a colossal amount of energy over 'natural' paper. A consignment of 1,000 supermarket bags weigh 140 lb if made of paper, but only 15 lb if plastic (so the plastic alternative saves 90% of the pollution from transportation). The paper bags take 18 cubic feet of storage space, plastic only 0.4; and the paper version takes three times as much energy, and ten times the

KEEP THE OCEANS CLEAN THIS CHRISTMAS

OUT WITH THE PLASTIC TAT	IN FROM THE COLD
■ Balloons	■ Real, live trees
■ Tinsel	■ Paper chains
■ Plastic baubles	■ Home-made decorations
■ Foil wrapping paper	■ Bad jokes in home-made crackers
■ Plastic food packaging	■ Gummed paper and string
■ Sticky tape	■ Newspaper wrapping
■ Plastic bags	■ Bags for life
■ Plastic straws	■ Paper straws
■ Throwaway plastic cups, plates, glasses, cutlery	■ Proper plates, glasses cups and cutlery
■ Shop-bought crackers and decorative tat	■ Home-baked presents

A perfect example of the failure to understand plastic appeared in a popular newspaper last Christmas. This advocates throwing out plastic decorations, thus adding to the potential pollution problem, and advocates purchasing biodegradable alternatives. These (like paper and cardboard) are far more energy-intensive and polluting in production, and (as they biodegrade) they release potent greenhouse gases back into the atmosphere.

cost of plastic, to produce. People love to jump on bandwagons, rather than face facts. A glass bottle takes twice or three times as much energy to manufacture compared with plastic and pollutes the air with five or six times as much greenhouse gas. Each time a glass bottle is washed it consumes energy, and it produces potential detergent pollution. A plastic bottle is less troublesome to produce, and (once shredded) can be used for new housing or streets, locking away its carbon content for eternity. The public believe that paper cups are a menace compared with traditional ceramic crockery. In fact, they are far less energy-consuming than making and recycling china. However, none of this is known to the public, so

Rather than dispose of plastic (or try to make it biodegradable, which just adds more CO_2 to the air and would threaten the world's water supplies) it should be re-purposed. This road (*top*) has been made using shredded plastic waste instead of limestone aggregate, and here is a new building (*bottom*) made using cement and plastic waste in a trial carried out by Bath University and Goa Engineering College.

keep it between ourselves. And, if you do discover that miracle microbe that can degrade all plastic, best not to announce it until you're old. No point in seeing society crumble away before your eyes, just because you wanted to make your mark in history. There are limits.

COMPUTER SAYS NO

The 1971 section on computers has come true with a vengeance. Computers now dominate our lives, algorithms take over many of the decisions taken on our behalf, and this digitisation of data has revolutionised society. Actually, that isn't true; it has revolutionised the way we work, but society has not yet dragged itself into the computer era. Television programmes still have coy little warnings about 'not being suitable for children' because of 'coarse language and sexual references', heedless of the fact that the little darlings have been on a porn channel at their friend's house, watching persons with improbable genitals humping each other with contrived exuberance.

Everybody now has access to everything, but society in general hasn't woken up to the fact. For instance, computer fraud is everywhere. If money is taken from your account and placed in someone else's, the bank's default position is that it's your fault. The whole point of the digitisation of data is that they know exactly where every transaction occurs, so they should be able to reclaim it with a click or two; but they never do. The bedrock of the computer revolution was to have a secret store of data on everyone. Much of the money behind Facebook came from the secret service, who wanted a single portal through which they could mine personal data on everybody. The story of Cambridge Analytica, founded in 2013 to exploit personal data and brought down in 2018 for exploiting personal data, is an object lesson. Like Facebook, they were set up to gather data and use it for commercial purposes. As I explained in the original *Nonscience*, that was the whole point of it all. But spoilsport journalists became wise to the situation and let the public in on the secret. To Experts everywhere, this was a disaster. Analytica was forced to close down, and Facebook started selling advertising to make money instead. Fortunately, there are so many other avenues where the public don't realise what's going on, so the scope for progress remains wide open. The public innocently imagine that the Internet is there to help them, whereas we all know that it exists for them to help us. Everything they do is carefully monitored and observed in detail by any authority that wants to know. Observers can turn on people's phones, and webcams, and microphones, any time they like. And it is getting better all the time. People used to warn of the era of Big Brother, invented by George Orwell in his novel *1984*. In his scary novel, Orwell promoted the idea that: 'Big Brother Is Watching You!'—and this was how the world of the future would be controlled. Well, it is much worse than that. It isn't just that Big Brother can watch everything the public do, for the modern world allows just about anyone to hack into just about anything and check up on just about anybody they choose at just about any time they want. Members of the gullible public make it so easy—many have actually installed devices like Alexa in their homes. Just think what this means: people can hack into those and follow every intimate word, each secret discussion, passionate encounter, burp, fart and shouted argument, any time they want. It's Orwell on steroids. The latest development is that Amazon, Apple, Google and the Zigbee Alliance are working together to make their devices all talk to each other. It's paradise! You won't even have to ask an Alexa device to listen in to something important—just ask a device that Alexa knows. Experts now monitor everything people do. Heaven!

Computer people are an illiterate lot. That's why a document or an image in your computer is called a 'file'. Of course it isn't. For 500 years a file has been something that you put documents and images inside, not the items themselves. But you have to go along with it and keep up to date with the latest software. Those working with computers compensate for their inadequacies by being boastful, and that always works. Experts keep reminding us that computers can do things the human brain cannot. That's hardly new: so can a stapler or a pair of scissors, and nobody thinks they're intelligent. There are many software systems now that carry out digital automation, and have cybernetic systems that adjust the results in the light of previous encounters. This is very useful, and (to make sure people realise just how special it is) it is now known as 'artificial intelligence' (AI). It has nothing whatever to do with intelligence. The AI myth has been perpetuated by the wild exaggerations of people like those who built 'Alpha Go', the first computer to solve the board game 'Go' by playing against itself. This isn't intelligence! Intelligence is finding answers to unforeseen problems—this is just playing a board game with rules set by people. The DeepMind project also fondly imagines it is dealing with the realities of intelligence. These people love to say that they are solving human intelligence so they can use that to solve everything else. They say they are aspiring towards The Singularity, when human brains and computers combine. That is never going to happen, and these folk know nothing whatever about human intelligence. Theirs will never be a 'theory of everything', only a theory about the little they know. It's the same with robotics—workers have objected to automation since the 18th century, and today's 'robot revolution' is just a continuation of the same process.

The public are incredibly excited to know that surgery is often carried out by robots. The robot doctor, say the reports, is better than a human surgeon can ever be. Now, we all know this is absurd. Robots must have a certain degree of autonomy to be defined as a robot, and the surgical robot doesn't.* It is actually a micromanipulator, which is controlled by the surgeon's own hands. But wait: how impressive would that seem to the uninformed public or the ignorant media person? To tell people their surgeons use micromanipulators to move their scalpels

---

* People are actually developing new robots for use in operating theatres. One idea is for them to undertake suturing, where neat, precision stitches are needed. This would be like a sewing-machine that steers itself. There is also a proposal for a small robot, like a self-propelled lawn-mower, which could navigate itself to find damaged heart tissue and automatically insert a valve, or whatever. These really are robots—but of course, the name has already been used. The PR department will have to find a way to jazz it up. I am sure they will.

The media are enthralled by the idea of a robot doctor. These devices are not robots at all, but are actually micromanipulators. They have existed since Queen Victoria was alive, but we must never say so. Calling them robot doctors is always good for headlines. They have introduced lots of new problems of their own, and patients sometimes suffer from this, but it's best to play down the fact.

doesn't sound remotely impressive—it almost makes the surgeon seem ineffective, if left to their own devices. Instead, we say: 'Your robot doctor will see you now' and everyone is massively impressed. That seems like the very latest sci-fi innovation, even though it's a barefaced lie (or a 'statement not wholly consistent with demonstrable factual determinants', as we prefer to put it). It isn't even new—there is a paper in the journals on this subject which sounds incredible, entitled 'A technic for the inoculation of bacteria and other substances into living cells'. This showed how you could use a micromanipulator to move a cell nucleus from one cell to another, and even to pick up single bacteria and inject them inside living cells. Isn't that incredible? And when was this paper published? It was back in 1911. The author was Marshall Barber of the University of Kansas who had invented the micromanipulator a decade earlier, when Queen Victoria was on the throne. Needless to say, nobody is interested in him, or his device. That's ancient history. And nobody is interested in micromanipulators today, either. But robots? A robot surgeon? That's the way to go.

Remember, people are most impressed by robots when they look like people.

A 'social robot' called Sophia appeared on the Jimmy Fallon television show in the United States, where he described this creation as 'basically alive'. It was claimed to recognise human emotions. Another, named Pepper, visited the *Financial Times* offices and was hailed as a visiting dignitary, while conspicuously failing to respond to the commands given by people. The truth? These are dolls. They are like children's toys, only larger and sufficiently anthropometric to encourage intelligent people to interact as though they were conversing with humans. They're still toys.

I mean, she's basically alive.
Is that what you're saying?

When Pepper the robot went to visit the managing editor of the *Financial Times*, Robert Shrimsley, their verbal interaction could not be sustained (*top*). Pepper the robot just beeped, like a toy. The manufacturer assured the FT that Pepper 'could understand emotions' and the gorgeous, gullible journalists actually believed that. In America, another robot called Sophia appeared with Jimmy Fallon (*bottom*). He was just as easily taken in by this full-sized doll and the aura of hype that came with it, suggesting it was 'basically alive'.

The essential difference between these robots or AI systems and humans is that they are constructed by technicians. Humans, like other forms of life, create themselves. You can observe real intelligence in single cells. If a cell in a filament of the alga *Antithamnion* is cut in half so the cell contents empty out into the surrounding water, the two adjacent cells cooperate to identify the problem, and then to rebuild the cell so it is restored to life. This is an unprecedented situation for the alga—if a filament had been broken in nature (being snapped in half by a rush of water, or trodden on by a dinosaur) then the two broken ends would be a long way from each other. The only place where they remain close together is on a microscope slide in a laboratory. Even though this is a unique situation, these simple plant cells work out how to fix the problem and mend their neighbour. That's intelligence!

There is a way to assess a computer, to see how close it is to human intelligence. A volunteer interacts with it, and if the conversation seems indistinguishable from what real people would say, then the computer has passed the test (it is called the Turing test, because Alan Turing coined the idea in 1950) proving that it matches human intelligence. To me it makes no sense. All you'd have to ask would be: 'Were you unpacked from a carton, or emerged from a womb?' to find out the truth. In any event, that's not intelligence. Based on my comparison with *Antithamnion*, I would propose a new test for computer ability. If someone claims to have made a computer as intelligent as a single cell of an alga (not as bright as a human, we'll forget that for the moment) then this is the Ford test. One Friday afternoon you set up three computers, connected on a network, and then smash the middle one with industrial cutters and a heavy hammer. Make sure the case is broken completely in half with the contents spilled out across the floor. Next, lock up the laboratory and go home for the weekend. On Monday morning, go back to see what happened. If the middle computer is fully restored to as-new condition and is working perfectly, then you have made something as intelligent as a microbe. If not, then you don't know what artificial intelligence is.

FUNDING FOR FOOD

The subject of food confuses the public just as much. In 1971 *Nonscience* predicted that 'expiry dates' made little sense (a few people are starting to say so now) and there are some fine examples. I recently saw a bottle of 'mountain pure' water that they claimed had been filtered through rocky strata for 250,000 years. After a quarter of a million years in the making, the water now had a date of expiry before Christmas. There are health scares over food all the time—even over water. The original book (on its final page) warned that water, taken to excess, could kill, and in 1983 the *Durand Express*, produced in Michigan, published an April Fool's spoof on the subject. They said the chemical substance that had been detected in the city's water supply was dihydrogen oxide (DHMO). Few people realised that dihydrogen oxide was water. Many others have reprised the same joke; indeed a 14-year-old high school pupil* named Nathan Zohner of Idaho Falls launched a petition to have DHMO banned. He called his project 'How Gullible Are We?'

Fashionism reigns supreme in the field of food. A prawn cocktail is now passé, like a Black Forest gateau. And the way we ensure food is hygienic is equally fickle. In America they wash chicken carcases in dilute hypochlorite (sometimes with monochloramine, which seems to work better) in order to eliminate food-poisoning germs. In the leafy avenues of middle-class Britain, the notion of 'chlorine-washed chicken' is considered completely unacceptable. It has the hallmark of impure food. This is curious—the same families buy 180,000 bags of salad leaves every year. They even eat slightly more than half of those. And every single bag of leaves is … chlorine-washed food. Confusion over our diet is widespread, even more now the Prime Minister has announced his obesity strategy. He lost weight after his brush with death through COVID-19 so now insists everybody else does so too. Because he cycles a lot, he thinks everybody else should. But cycling for an hour burns 300 Calories; you can eliminate that more easily by not eating a salad sandwich. It is a fact that, if someone resolves to lose weight, all they have to do is stop eating so much food. Their expenditure is immediately reduced, and so is time spent in the supermarket. From the moment the decision is taken, wasted time and super-fluous expenditure are both reduced. Experts all know that, which is why they're slim. However, the public are not permitted to see it like that. In practice, if they decide to slim down, the first thing they are told to do is go out and buy expensive food for their diet. Currently, in Britain alone, this is an industry worth £20,000,000,000 (twenty billion pounds) and in America it tops $70,000,000,000 (seventy billion dollars). This is breathtaking—the British total is almost half as much as the nation spends on defence. Yet, if only they realised the fact, every penny is a waste of money. You should stop spending so much on food, not start spending more!

Many foods are simply dismissed as dangerous, while others are extolled as healthy. The one dish you must never eat is the good old traditional fry-up. The English Breakfast—egg, bacon, fried bread, tomatoes—is the kiss of death. Everybody knows that. On the other hand, you won't find a health-food restaurant in the world that doesn't serve quiche. It even gave rise to that macho collo-quialism: 'Real men don't eat quiche.' In fact, the ingredients of the quiche are exactly the same as the fried breakfast. The egg is in the filling, with the (chopped) bacon, while the fat and carbohydrate of the fried bread are in the pastry case of the flan.†

---

* Pupil? That should be 'student', unless they are already being called 'Professor'. It's only a matter of time.
† Remember to call this flan a quiche. Quiche is trendy and healthy. Flan is old-fashioned and common. Do keep up!

Quiche recipes also add grated cheese and thick cream, meaning that the quiche can be far more laden with saturated fat and Calories than the fried breakfast. And the fry-up had tomato, which gives it a healthier profile than the quiche! There is a similar argument against fast food, like a hamburger* or a kebab. Both are ancient, traditional foods. Jesus ate kebabs. And hamburgers are a thousand years old at least. The blend of vegetables (salad or onions) and ground meat with a carbohydrate wrapper is an excellent combination of nutritious foods. The problem is the speed with which people eat them; not the speed with which they're served.

Currently, people are taught that they need about 2,000 Calories per day for women and 2,500 for men. I have shown that the actual totals are more like 1,600 Calories a day for women, and 1,800 for men. The standard figures date back decades, when people had cold bedrooms and children walked to school. People worked in their homes and factories, instead of (as happens now) sitting in a warm house, sitting in a heated car, and then being seated at work all day in the warm. Children are driven a mile or two in a heated car to their air-conditioned school, and adults are accustomed to crouching in a stifling vehicle breathing toxic fumes on a crowded highway for a total of at least one working day per week, before spending all day in a heated environment with the windows shut. This confined and controlled existence has slashed the amount of food we should eat, because our energy needs are tiny compared with our forebears. Instead, people still eat the same amount of food, and wonder why they often get so fat. They really need far, far less, and my (roughly) 1,700-Calorie figure is closer to reality. Nobody should eat meals as our parents did. A single meal provides the sustenance we need for an entire day; instead, some of us still eat three. Experts never say so, however.

The kind of regime that people adopt will give you ample scope for fame and power. The Atkins Diet was followed slavishly by millions. Dr Robert Atkins promoted it all his life. Although his followers never knew it, he was prone to heart attacks, had high blood pressure, and suffered from congestive heart failure. Equally popular was jogging, promulgated by Jim Fixx who wrote a book called *The Complete Book of Running*. He collapsed and died aged 52, while jogging. There is a great vogue (summarised by 'no pain, no gain') that suggests you have to hurt yourself to obtain benefits to health. All this is nonsense. If you're hurting, it is because your body is telling you to slow down.

Experts must always remain in command of what people believe. For example, you should *never* allow people to think they're fat. Instead, they are 'plus-size', 'heavy', or 'large' and Experts have now introduced the idea that, if you react to someone because of their excessive weight, you are guilty of 'fat-shaming'. People are currently 'nicotine-shamed', which is a curious reflection on recent history.

---

* Why do people keep calling this a 'beefburger'? The hamburger is named after the German city of Hamburg, from where they originated. It has nothing to do with ham. They're all beef.

There are popular television programmes (like the bake-off shows) which demonstrate how to make cakes people can eat like gluttons. The main dietary components that are doing us harm are excessive fats, sugars, and carbohydrates; and it is sugar, flour, and fat (the major unhealthy foods we eat) which are the main ingredients of the fattening cakes and pies on these baking shows. Why, 'death by chocolate' is heralded with pleasure all around the world. There was once a tradition of drunk driving, exemplified by the words: 'One for the road!' at the end of a night's drinking. They could have called it 'death by drunkard' but I don't think that caught on. Bake-off programmes have a significant mortality rate and it won't be long before some tragically bereaved family will sue the producers. These programmes kill people. Legally, every show should have a health warning. As it is, these appalling series ride high in the charts and I am waiting for the 'Great Tobacco Smoking Challenge' or the 'Blindfold Railway Crossing Elimination Game'.

If you do overeat spectacularly, you will become obese. But Experts always have a list of excuses to prevent you realising why you're fat. The simplistic answer of 'eating too much food' is kept hidden. Cooking odours are often blamed; so is Distracted Eating, when people nibble without realising it; the abundance of nourishing food is also blamed, so is fast food; Portion Distortion (where individual serving sizes have increased) is another culprit, and so are the giant-size food packages available in stores. It may be that you have a slow metabolism, are big-boned, or have an inherited tendency to put on weight. You see? Those factors are to blame, never the person who eats. Experts know that there are no subtle, genetic, physiological obstacles that make people fat. We know this because the one single, guaranteed intervention, invoked as a last resort, is the gastric band. This is a kind of zip-tie which makes the stomach so small it can no longer hold large amounts of food. It has one major effect: people thus treated simply cannot hold as much as they did before. They don't have to bother with willpower or decision-making, for the plastic band stops them eating. And they *always* lose weight. So, none of the diet advice, speciality dieting, branded diet supplements, fad diets, etc., are guaranteed to succeed—but physically preventing people from eating will always work. Fat people are fat because they eat too much. Just never tell people about it; this is our little secret and there is too much money at risk if the public ever found out.

LUNGS OF THE WORLD, NOT

Currently, top of the chart is the destruction of the Amazon rainforests. The news media are pumping out scare stories about the loss of our vital oxygen supply. People are buying oxygen cylinders to ward off suffocation. Millions are concerned that land clearance is carried out by money-grubbing speculators heedless of the damage they cause, and that our air will become weaker and end up unbreathable if the burning doesn't stop. They all know that the Amazon is the 'lungs of the world' and the record levels of destruction are going to harm us all. This is a huge money-making enterprise; there are over 350 million websites on the subject, with half a million discussing the 'lungs of the earth' problem. Experts are pulling in hundreds of millions of dollars to fund these concerns. It is a massive growth industry.

Of course, not a word of this is right. The Amazon isn't anybody's lungs, and it never has been. And it isn't only the Amazon that is being burned.

Rainforests are being destroyed all around the world. In Borneo vast swathes are being cleared, as they are across much of Indonesia. Once in Yangon (previously Rangoon), the capital of Myanmar (formerly Burma), I saw the dazzling city view rapidly vanish into acrid smoke from rainforest burning 500 km away. Much of SE Asia is smothered in smoke from rainforest destruction. Most worrying is the Amazon, simply because that rainforest is so rich in wildlife.

In the space of 20 years, tribes along the Amazon have acquired mobile phones and satellite television while their dugout canoes have been replaced by skiffs with outboard motors. Tourists love to see people living primitive lives and resent the impact of the modern world. We'd rather they stayed primitive. The people of Amazonia look at us and see how wealthy we became by ripping up our environment and killing off our wildlife, and feel they might like to catch up.

Our Fenland in East Anglia was once a kind of rich and verdant rainforest. Now it is flat farmland with serried rows of crops, and almost all of the rich peaty soil has been oxidised back to CO_2. Soon it will be gone. That's fine, of course, because it's us. It is only when someone else tries to follow our example that the trouble starts.

The secret is that the Amazon has never given us any of our oxygen. The oxygen we breathe was liberated by microbes millions and billions of years ago, and the huge strata of limestone we see are an example of where the carbon is stored. That's what gave us a breathable atmosphere, and the amount of oxygen hasn't changed since. If you burnt everything in the world, it would have little effect on our oxygen levels. No amount of forest burning could affect the way we breathe!

It is true that forest trees give out oxygen, but that happens only when the sun shines. At night trees respire as we do, taking in oxygen and giving out CO_2. As each tree grows, it locks away a little carbon each day until, by the time it is mature, the average tree has captured about a tonne of carbon. It will have given out a similar amount of oxygen. But (and this is the point nobody has recognised) when the tree dies and decomposes, all that stored carbon is released back into the air as CO_2 and every atom of oxygen that the tree liberated as it grew is taken in again by the microbes that are degrading its remains. Through its completed life cycle, from the germinating seed to the final decomposition of

'LUNGS OF THE EARTH' ON FIRE   TOI

Amazon: World's largest rainforest

BRAZIL

- Located in South America
- **Covers 5.5 million sq. kms** of land
- Spreads across **9** nations
- **Brazil accounts for 60% of Amazon rainforest**
- Produces **20%** of the oxygen in Earth's atmosphere
- Home to **3 million** species of flora and fauna; **1 million** indigenous people

Aug 23 fire map
by NASA

Area lost to deforestation (in Km²)

Year	Area lost
2013	5,891
2014	5,012
2015	6,207
2016	7,893
2017	6,947
2018*	7,900

#PrayForAmazonia

THE AMAZON RAINFOREST **"THE LUNGS OF OUR PLANET"** PRODUCES MORE THAN 20% OF THE WORLD'S OXYGEN AND IS IT **ON FIRE.**

AMAZON RAINFOREST —IN NUMBERS—

73,000 Approx number of fires since January 2019

4,100 miles of winding rivers

1/10 known species on Earth live in the rainforest

Home to 1 milion indigenous people and 3 million plant and animal species

20% of the earth's oxygen is produced here

Spans across 8 rapidly developing countries; Brazil, Bolivia, Pweru, Ecuador, Colombia, Venezuela, Guyana, Suriname and French Guiana

9 If it was a country it would be the ninth largest in the world

1.4 billion acres of dense forests, half of the planet's remaining tropical forests

2.6 million square miles in the Amazon basin, about 40 percent of South America

The rain forests contain 90 to 140 billion metric tons of carbon

Media reports around the world remind us that 20% of our oxygen comes from the Amazon. The lungs are on fire, according to the *Times of India* (*top*) while Express Newspapers remind us that this is where our oxygen originates (*bottom*). This must be true because everybody says so. An ungrammatical statement from Pray For Amazonia (*middle*) emphasises the urgency by finding that the real figure is more than 20%.

Soya plantations reaching towards the horizon have replaced the lush verdant vegetation after slash-and-burn clearance. Millions of mature trees and countless wild animals have been senselessly destroyed, and the indigenous population have been driven out or killed by rich city speculators. Actually, this is in the English fens. We certainly don't want anybody else following our example.

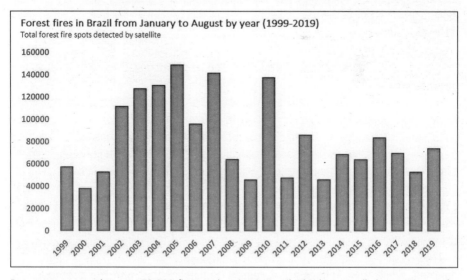

Forest fires in Brazil from January to August by year (1999-2019)
Total forest fire spots detected by satellite

Recent reports say that over 70,000 fires are burning in Brazil. This has actually been going on for centuries, as peasant farmers try to keep up with what we have already done to our environment in our quest to become wealthy. There are thousands more rainforests burning all around the tropics, but we don't mention those. The Amazon is where the money is, so that's the target of Expert attention.

Wet with Amazon humidity and clinging onto a branch with matted hair and rheumy eyes, the author encounters a two-toed sloth in the Amazon rainforest. Until we in Britain decimated our own forest environment, you would have seen our own bears and wolves. But becoming wealthy meant destroying our wildlife, so that's what we did. Why the cheeky Brazilians are trying to catch up is a mystery.

the last scrap of rotten wood, each forest tree bequeaths nothing whatever to the atmosphere.

But don't tell anyone I said so. We love our pervasive myths, and there are Experts with large homes to run and with kids at private school who would lose out if the truth were revealed. The people who run those charities need their fortunes (the CEO of the American branch of the World Wildlife Fund earns more than the American president) and it would be churlish to spoil their fun.

As the original *Nonscience* foretold, the world of the Expert is now run with colossal sums of money for projects nobody can understand, organised by people with an aura of untouchable superiority who speak a language that no-one can comprehend.*

These are the Experts who cannot be questioned; whose word cannot be doubted; while their actions are accepted unquestionably, and the consequences simply have to be endured by the public. Few of their projects are genuine, and

---

* This includes them. Experts rarely have a clue what they're saying.

many of their fundamental beliefs are false. Even the most prestigious authorities in the land can cheerfully misappropriate other people's ideas or steal someone else's intellectual property. The most self-evident absurdities can be embraced, the greatest distortions sanctified, the most wild and expensive dreams dignified … this is a world of massive opulence and extravagance where the most costly devices in the world are put to the most trivial of purposes (assuming you can find somebody who knows how they work). In any other field, people would dismiss these people as a bunch of dangerous opportunists—but this is Nonscience. It's the new religion.

Just as they once trusted priests, people have unquestioning faith in Experts, and—when we explain how complex a subject is, or how long it will take to investigate, or why we need colossal sums of unaccountable money—the public meekly accept our word for it. But there are warning signs that this might change, and the COVID-19 pandemic is a case in point. The Chinese quickly had their scientists on the case. The disease was recognised late December 2019, and by early January 2020 they had already sequenced the entire genome and released it to the world (which enabled us to start testing). By April there were about 120 vaccines under development and in the United Kingdom—where news of these other vaccines wasn't mentioned—we were promising to develop a new vaccine within months and have it on the market in record time. This is such a dangerous trend! We need to nip this in the bud—nothing in Nonscience should ever be done quickly.

When we face emergencies, all the bureaucracy that Experts adore is forgotten. It happened in WWII, when the Germans pioneered the world's first single-seat jet fighter, the Heinkel 162 Volksjäger. Its design was first agreed on 25 September 1944 and it made its maiden flight on 6 December 1944, less than 90 days later. Experts would want at least 90 months of preliminary discussions before planning even started. It was much the same story in 1957 when there was a global pandemic of H2N2 influenza, Asian 'flu. It was a virologist from the University of Chicago, Maurice Hilleman, who read reports about the outbreak in Hong Kong and developed a world-beating vaccine in four months. Sixteen weeks! It's ridiculous. Hilleman joined the staff of the Merck company to study vaccination and he personally created vaccines against mumps and measles, meningitis and pneumonia, hepatitis A and B, and a host of others. He worked on over 40 different vaccinations and has saved millions of lives. This is just what we don't want! You need dozens of committees writing endless reports and consuming vast sums of money. Whatever happens, never let scientists like Maurice Hilleman loose on the case.

We need to pour scorn on the opposition. China (which has one of the most pernicious governments in the world) hid the outbreak of coronavirus for weeks. They refused to admit anything for far too long. Keep reminding the public about this. Don't ever reveal that the first case of BSE in Britain was in Easter 1985 yet the government managed to conceal the truth and issued bland reassurances until finally admitting the outbreak of vCJD in humans in March 1996*. Their silence could have resulted in millions of British people dying of this crippling brain disease, and it is only through good fortune than it didn't develop as widely as we feared. These things must be kept secret; transparency is the enemy of Nonscience. Similarly, we keep saying that the incidence of COVID-19 in China may have been 50% higher than they admitted, but quickly gloss over the fact that our own official figures were probably ten times too small.

Science teaches us that you stop an epidemic by closing borders, tracing contacts, and introducing rigorous quarantine. Asian countries did this, and they were copied by New Zealand. They are now virtually back to normal.

We stopped tracing contacts and allowed people in by air, loosened lockdown, and abandoned testing for months. Sampling kits were supplied without a return address, or with swabs that wouldn't fit their test-tubes; so Experts compensated by emphasising the large number of tests being carried out (and included tests still in the pipeline). Testing is useful only when people are tested regularly, and sometimes we reached about 90,000 tests per day. At that rate it would take two years to test everyone in Britain. Instead of using tried-and-tested software, we spent over £12,000,000 on a doomed NHS app using hirelings recruited by Serco at a cost of £45,000,000 who would ring contacts in the hope they might self-isolate. The app didn't work on 96% of iPhones, and within days fraudsters were ringing round telling random members of the public to self-isolate (but first to send them a diagnosis fee). We are currently looking at apps like those that have been in use elsewhere since January, which will take many months to introduce, by which time thousands more people will have died. And meanwhile those avuncular government officials will simply keep smiling and telling the public this is all part of their grand plan.

COVID-19 has changed the way Experts interact with politicians—and there are enormous differences between the two. Experts are brainy and successful as they strive for power by exerting an aura of unquestioned authority, so people believe what they say. They need meticulous training. Indeed, you've just read an entire book explaining this. Politicians just become important overnight, with not a whiff of experience. Most have failed in their careers, which is why they are available to stand for election. Typically, they are endearingly dim-witted social misfits who can make up ridiculous rules and change their minds whenever they want, whereas Experts always know what they're after, and never change their minds about anything. Experts use complexitational obfuscationary obscuranticism to impress, while politicians don't know complicated words. Their slogans are three words long (like 'get Brexit done') because most can't concentrate on any more. Boris Johnson vaguely remembers some Greek mythology from school, so he adds some to his stumbling speeches, while the Home Secretary, Priti Patel,

---

* There was an absolutely excellent book written about the facts of BSE, but I can't remember the details now.

recently reported that the government had carried out 'three hundred thousand and thirty-four, nine hundred and seventy-four thousand' virus tests and didn't bat an eyelid.

The SPI-M committee (page xviii) concluded on 10 February 2020 that there definitely was 'sustained transmission' of the virus in Britain; two weeks later Public Health England assured care homes that there was 'currently no transmission'. Thousands died as a result, and a Commons committee said they had been 'thrown to the wolves' through 'reckless' and 'appalling' errors by government. What a negative attitude! This will save the exchequer billions of pounds that would have been paid out in pensions and, as Professor Robert Dingwall said on 2 August, they were due to die anyway. That same month the government promised that everyone in care homes would be tested but broke the promise within weeks, and we saw on page 54 how Mr Johnson and Mr Hancock state untruths. People call this 'lying through their teeth' but, whether a statement is true can be a matter of debate, and if they do so through their teeth, or through any other part of the body, it is not for us to judge.

As we saw on page xviii, the Prime Minister's mentor, Mr Dominic Cummings, drove his family 550 miles across Britain, scattering COVID-19 virus wherever they went, to spend time with his family. This was what millions of other people wanted to do but were prevented by law. Mr Cummings kept this secret (even hiding it from the Prime Minister). His simple-minded wife wrote a deceitful article pretending that he 'lay doggo' until emerging from quarantine 'into the almost comical uncertainty of London lockdown' but lay people saw them in the North of England, and eventually there was a press conference at 10 Downing Street. Mr Cummings, who doesn't think very fast, read a stumbling statement that justified it all, and then fielded bland questions with inarticulate, but rehearsed, responses that emphasised his need to drive north in case he needed childcare for his dear little son, Alexander Cedd (appropriately, named after a saint who died in the pandemic of 664 AD). He claimed regular child-minders weren't available, so asking his family for help was his only option. What no-one thought to raise was that Mr Cummings, as the Prime Minister's adviser, is the most influential and high-placed individual anywhere in Britain, so all he had to do was lift the telephone and say: 'I may need child care if we both become ill,' and someone would have said, 'Right away, Mr Cummings.' Having to wander across Britain asking your family for help is what mere mortals do, not exalted overlords like Mr Cummings. Arranging a side-trip to the family's favourite beauty spot on his wife's birthday was such a romantic idea; everyone knew such a jaunt was illegal, and so Mr Cummings explained that he risked driving his vulnerable family to Barnard Castle only to see if he was still safe. One of my editors remarked: 'Little shits; they should be sacked.' Rude comments like that cannot be justified: they are not particularly little, and—if someone is willing to employ people like that—it is not for outsiders to question their judgement. The Prime Minister ruled: 'I think that was sensible and defensible and I understand it … he behaved responsibly and correctly and with a view to defeating the virus and stopping the spread.' He ended further enquiry by saying he had 'drawn a line under it' so it was 'time to move on'.*

It doesn't only happen here. The American President announced in February 2020 that: 'the Democrats are politicising the coronavirus … this is their new hoax.' Dr Jane Appleby of San Antonio, Texas, reports a patient who went out with an infected person to see what happened and died in hospital saying: 'But I thought this was a hoax! It isn't.' In June a party was held in Dallas, Texas, by the Green family who accepted it was a hoax, and 14 people became infected. One died. Mr Trump cleverly proposed an injection of bleach or a beam of light shone through the body as a cure. He insisted: 'It's one person coming in from China and we have it under control. It's going to be just fine. China has been working very hard to contain the coronavirus, and the United States greatly appreciates their efforts and transparency. It will all work out well.' When announcing the results of his own coronavirus tests on 21 May 2020, he stated: 'I tested very positively in another sense so—this morning. Yeah. I tested positively toward negative, right. So, I tested perfectly this morning—meaning I tested negative.' Ms Patel would have been so proud.

Our former Prime Minister was wrong to say there is no 'magic money tree'; there are billions to be had and a government can borrow all they want, without much interest to pay. This is always announced as the Chancellor 'giving' money for a scheme. In reality the British people are being saddled with yet more debt. The total must be about £4,000,000,000,000 (four trillion pounds) which is more than our entire economy is worth. The debt does not rest on the unemployed or impecunious pensioners, who are about half the population, so the money owed is actually £120,000 for each working person.

If you raised tax by £2 a week it would take 1,200 years to repay but the government doesn't mind, because they won't be in power. They have a unique timescale. To the public, 'long term' means over the next few centuries. To an Expert, it means the time before you retire. To politicians, it means by the next election—five years, tops.

The government said they would donate £30,000,000,000 (thirty billion pounds) to support jobs, though much of it had already been earmarked. For example, they promised £400,000,000 (four hundred million pounds) for trainees and apprentices, but half of that had been announced ages ago.

In one of my favourite shopping centres in Miami, Aventura Mall, you would find 'COVID-19 Essentials', a store selling disinfectants, sanitisers, sterilising lamps, masks, shoe-cover dispensers … some complained it was a cash-grab. Of course it was! That's how free enterprise works. Our Prime Minister graciously permitted people to go shopping again in July, if wearing facemasks. They hadn't needed

---

* Goodness. You'd think he had been reading this book.

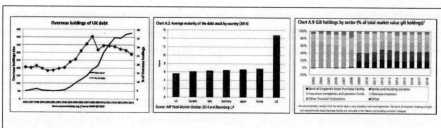

UK External Debt

It is the total amount that people in a country owe to the rest of the world. It includes both **government debt** and private sector debt. The vast majority of this is liabilities by the banking and finance sector. UK banks external liabilities amount to over £4,000bn.
23 May 2020

Currently the British people have been saddled with debts of over four trillion pounds. This graph is from the economicshelp.org website, and it means that each working individual is in hock for £120,000.

masks from January till July, but now they suddenly did. I particularly liked the advice of Professor David Heymann, who said: 'Wearing a mask must be consistent; it's not on to wear a mask and then decide to take it off ... to eat a meal; it must be worn full-time.' By 8 July, isolation had reduced infections to an all-time low, so Boris Johnson announced his relaxation of the rules. Within three weeks, precisely as the science predicted, the numbers zoomed up again, so the Prime Minister cancelled everything he'd promised, and establishments busily preparing to open were shut down again. It was like stopping children playing on the railway line because some were killed, and then—after a few months without a fatality— telling them it's alright to go and play on the tracks once more and then being surprised when the 7:17 to Sevenoaks severs some little lad's limb.

Politicians control the facts they release, and they wouldn't disclose levels of infection, so 600,000 people jetted off to Spain on the busiest holiday weekend in July, only to be told (with four hours' notice) that they'd have to be quarantined

The public think of time in a realistic way that affects the future of society. Experts relate it to their lifetimes. Politicians look only to the next election, when all that matters is that they persuade people to vote for them, and not those other ghastly nominees.

Sector of society	Short term	Medium term	Long term
PUBLIC	Months	Decades	Centuries
EXPERT	Weeks	Months	Decades
POLITICIAN	Days	Weeks	Months

when they returned. Politicians can do this. They just revel in their unaccount-able power. For weeks, the British government refused to provide details of who was infected, and where they were, to local authorities who desperately needed to know, while England was recording more deaths from COVID-19 than any European country. Similarly, in America, the government watched numbers soar to record levels, so the President argued they should carry out fewer tests and thus reduce the number of known cases. He didn't like the way the facts were analysed by the Centers for Disease Control (CDC) so they invented the Department of Health and Human Services, and made sure the data went there instead. As a result of the confusion, estimates of mortality range from 1 person per thousand to 250. Huge numbers are permanently disabled by what they call 'long-covid', but nobody counts those.

The rules and regulations caused a jumble of confusion for everyone across Britain. Small shops and expansive garden centres had to close while the rival supermarkets remained open. After all the vast extra profits they made, I think supermarkets should pay a windfall tax. In July it was perfectly legal to trim a man's beard, but not to thread a woman's eyebrows. You had to wear masks to enter a takeaway, but not to attend a meeting of Cabinet. Nobody in Britain had to wear a mask in the gym, where you're sweating and bouncing about, but you did to sit quietly on the train, where you'd be turned away if there were too many passengers aboard, while people crowded into busy pubs and clubs with not a mask in sight. The government banned all theatre performances, because people sit in rows, and all cruises, because passengers were in a confined environment; but allowed people to travel by plane, where they were both confined and sitting in rows. Politicians don't usually go on cruises, nor are they encultured enough to visit the theatre—but they do adore plane travel. Over 90% of people understood the rules in March 2020; by August fewer than half could work them out. Boris Johnson defined these results as a 'massive success'.

It has been suggested that I should go into politics, but it never appealed to me. Were I the Minister for Science, I'd go to the despatch box and present my govern-ment's radical new policy, full of fresh ideas. Then the shadow Minister would rise and she* would respond on behalf of the opposition.

Then I'd likely turn to my side of the House and say: 'Oh, her idea is much better than mine! Let's do that instead.' So I am sure my political career would be noteworthy, but it wouldn't have lasted long. Sometimes people have asked me whether my politics lean towards left or right. I think of my politics as being 'up' rather than 'down'.

Boris Johnson admitted they'd handled it wrongly from the start because they 'didn't understand it'. Other European countries acted differently. Norway and

---

* I always say 'she' to surprise people, because they automatically assume that a science minister must be a man. Sorry, a male.

Protesters in London march to protect their right to perpetuate the pandemic. Experts need this to ensure grant money keeps pouring in, and it reminds us all that this is a democracy. If it is your grandma, then you have every right to infect her and (from a safe distance) watch her die. No mere outsider has the right to interfere with your choice.

Denmark closed their borders to foreigners less than two weeks after their first cases. They recorded 40 and 87 deaths per million people, respectively, while our figure was 438 per million, the United States 421. Imagine if we had trace and isolate imposed, like some wretched communist country. Consider two I've visited: Vietnam closed schools, while all foreign visitors had to go to a residential quarantine centre for two weeks. When there were outbreaks in Ha Loi and Son Loi, both settlements were sealed off until two weeks had passed without incident. Similarly, Cuba closed their borders and monitored the population, testing and quarantining contacts. Every patient who tested positive was put into an isolation hospital. Can you imagine such interference with personal liberty? It's a disgrace.*

There is a growing movement of people who don't believe in vaccines. Dr Andrew Wakefield claimed to prove a connection between MMR and autism, but we know that's wrong because Japan doesn't use the MMR vaccine yet has higher levels of autism than countries that do. Even so, one-fifth of people in Britain and

---

* Cuba had about 2,500 cases with fewer than 100 deaths; Vietnam reported less than 400 cases, with no deaths. The way these people interfere with personal freedom; it would never be tolerated in the Free World, where people have the right to spread infections as much as they want.

America now say they would not vaccinate against the virus. This is splendid! Smallpox vaccination was insisted upon in the days when people were less inclined to extol their civil rights, and by 1975 smallpox was finished* and a field of highly profitable research was wiped out.

Then there was polio, which paralysed (or killed) millions of people within living memory. In July 2020 there were just 39 cases reported in the whole world. Because of vaccination, it is long gone from the Americas and Europe, and the vaccine has saved over 17,000,000 lives. Measles would be next. It is one of the most highly infectious viruses. After widespread vaccination, the United States was declared measles-free in 2000, and Britain followed in 2017. Then the anti-vaxxers came to prominence, and measles has returned. 1,000 children used to die from it in Britain each year, and vaccination brought that down to 20 by 1978 and to zero by 1996. Yet children die from it now, because of a refusal to accept the vaccine. Nobody has died of measles vaccination, though about 223,000 children die each year from road accidents, and another 127,000 from air pollution. Being vaccinated is one of the safest things we can do; much safer than driving, eating, or staying home.

Here is some private advice about COVID-19, dear reader, to keep you safe. First, never use disposable plastic gloves! Once contaminated, they deposit pathogens on everything you touch, while the air trapped inside encourages bacteria on your hands to multiply. Use cotton gloves, which don't transfer germs. For push-buttons or touchscreens, always use your knuckle, not a finger. Never use buffet tongs. One person using tongs can infect thousands but picking up an item by hand infects nobody. Contaminated clothing or other items (like mail) can go into a chamber (or an oven) at 90 degrees Celcius to be completely coronavirus-free. If you wear a facemask, bear in mind that a Norwegian study found that 200,000 people would have to wear masks for a week before one single case of infection was saved. Remember there is no mask with holes anywhere near as small as the virus and discarded masks cause world-wide pollution. Never admit COVID-19 patients to any normal hospital. Their isolation wards should be designed to have negative pressure air-conditioning, with staff wearing positive-pressure bonnets to prevent viruses entering. Drivers, from bus-drivers to pilots, should have positive-pressure cockpits, to keep out germs, and aircraft could use their emergency dropdown masks to feed pure air to passengers instead of oxygen. Short-wave ultraviolet lamps can quickly sterilize an area from viruses. Washing can exacerbate dermatitis, so rinse your hands in a weak solution of hypochlorite— and use a cloth dampened with the weak bleach solution to wipe down articles

---

* Well, in 1978 Mrs Janet Parker died of it in Birmingham but she caught it from a research laboratory, so it doesn't count, and Experts prefer not to talk about that. Some spoilsport insisted that biohazard regulations were introduced so it doesn't happen so often now, though there were 1,120 incidents involving bacteria, viruses, and toxins in American laboratories between 2008 and 2012. I hear there may have been an incident recently in Wuhan, China.

(like shopping) to ensure they're virus-free. The Prime Minister has cheerfully said we might be back to near-normal by Christmas. I hear that the disease is more often symptomless and milder these days but, believe me, until we have a vaccine (think of Russia) COVID-19 isn't going anywhere.

Will there be a problem with a loss of Chinese funding for British universities? The Chinese love to send their youngsters to study in Britain—over 120,000 came last year, bringing with them £1,700,000,000 (£1.7 billion) in funds. The Australian Strategic Policy Institute has researched the important connections between British academia and China. Their National University of Defence Technology has links with Cambridge University to produce tomorrow's supercomputers. The Chinese say: 'This will greatly enhance our nation's strength in the fields of national defence, communications and higher-precision navigation' and think how useful that will be in their military activities in the South China Sea. It also helps develop weapons, which is why the US government blacklisted the project. Manchester receives funds from an organisation producing Chinese intercontinental missiles, which is highly profitable, while Imperial College has connections with the Harbin Institute of Technology in China, which carries out secret weapons research, while the University of Electronic Science and Technology in China sends research-ers to the University of Glasgow to develop the surveillance equipment used in Xinjiang to monitor those Muslim Uyghur people you may have read about.

One of the greatest dangers is that students (and pupils) didn't have to go to classes. Crowding thirty children of the same age into a single schoolroom, day after day, with a middle-aged teacher, is mediaeval. With COVID-19 keeping youngsters from school, they have discovered their freedom to learn, and we can't have that. Let's get them back into routine as quickly as possible. Similarly, college students have been discovering how to live freely in the real world. Most of the people they will eventually consult don't need degrees anyway: life coaches, marketing consultants, fitness trainers, computer engineers, counsellors, cleaners, investment consultants, web designers, gardeners, printers, inventors, therapists, car mechanics, builders, insurance brokers, designers, publicity agents, mentors, business advisers, plumbers, graphic designers, IT consultants, electricians, nutritionists, even estate agents … they don't need degrees to earn their very sat-isfactory incomes. Students may realise that much of what they're taught should have been instilled in school, where teaching is still divorced from the realities of the outside world. It's a cardinal rule that you don't learn how to live your life, but you are taught about out-of-date script-writers (like Shakespeare) and irrelevancies (like algebra), while you won't learn about how your body works, or how to cook or run your flat, how to think, or how you handle finance or the law. Students are discovering that they were wasting their time at a college demanding large sums of money simply to make them conform, all stretched out over three years. People complain that there's no free speech in universities, but there never was free speech. Like everything else at a modern university, it is very expensive.

It is the duty of every Expert to maximise the money, to stretch out the time, to multiply the bureaucracy, and to keep the truth hidden at all costs. That's why this book was so important when it was first published. Experts needed a helping hand to launch this global enterprise, and few of them would have guessed that so many could ever have garnered such vast resources. Can society afford all this? Of course not. Are the results worth it? Well, results are hard to find, if you do find them they won't mean anything to you, and those that do emerge can never be value for money. Should we question how our money is spent? Always. Though perhaps not in this case.

Experts everywhere now have the stranglehold they sought, and they can ride rough-shod over any opposition. This is the ultimate triumph of Nonscience— which is why the contents of this book must never be revealed to members of the Lay Public. Now you can see why outsiders were instructed not to look inside. To them it's a closed book.

And that is how it must stay. Promise?

CHAPTER 12

Can You Make the Grade?

TRY THIS EXPERT TEST ...

As a post-indoctrinational retroassessmentary service to readers, the author has magnanimously prepared a standard examination paper that you may care to sit. Rascally Lay People may flit through it in a matter of minutes; True Experts will, of course, take it in their stride over a period nearer an hour. It is all self-explanatory, and at its end you will have a profound (and, it is hoped, salutary) insight into your own chances of success in the exciting new world of Nonscience.

EXPERTISTICAL REORIENTATIONISTIC ASSESSMENT EXAMINATION PAPER

(to be sat during a hot, humid afternoon in July)

All candidates are to attempt seven questions from Section A, four questions from Section D, five questions from Sections B and C. Questions are to be answered in numerical order, sections to be completed in alphabetic sequence and *not* in the form deduced from the previous sentence. No other questions are to be attempted. Write name, age, address, nationality of maternal grandfather and blood group in indicated positions. Read all questions carefully before commencing paper. Time allowed 3½ hours (no candidate allowed to leave room before this time has elapsed). Start now:

Name and personal details:

....

....

....

....

SECTION A: *qu*FN Assessment ('Fashionism Test')

(1) Is your chosen topic for research mentioned enthusiastically in the press at the moment?

yes ☐   no ☐   occasionally ☐   frequently ☐

(2) How many years have elapsed since you first saw it mentioned in the press or on TV as a topical issue? Subtract total from 10 and mark it in the space provided. ☐

(3) Based on present trends, what do you consider to be the chances that it will be headline news in about 2 years' time?

excellent ☐   fair ☐   poor ☐

(4) How many Key Phrases does your Preliminary Account of Research contain? Multiply number by 1·72 and note it in the space provided. ☐

(5) Look carefully at the following terms. Which seems to you to have greatest importance in selecting a research project?

    a) altruism   b) social benefits   c) fashionism

    d) financial security   e) worthwhileness

Place an octagon around the letter indicating the term of your choice:

    a   b   c   d   e

(6) Look carefully at the following topics. Which seems to be the most propitious for a research project at present?

    a) world famine   b) ecology   c) prostaglandins

    d) the laser   e) nuclear fallout   f) drought

Add serifs to the letter adjacent to the topic you have selected:

    a   b   c   d   e   f

(7) Which is it best to gain in the course of your professional activities?

    a) membership of the Athenaeum

    b) a DSc (in addition to your PhD)

    c) Honorary Graduateship of the University of the Pacific

    d) a guest appearance on the *David Frost Show*

    e) the McPherson Gold Medal Essay Award

    f) a series contract for *Play School/Sesame Street*

Select two and mark them in the spherocirculoids provided.

(8) Pick five terms that are of supreme importance to you from the following list:

    a) rock and roll   b) chastity   c) drug law liberalisation

    d) honesty   e) integrity   f) hot pants

    g) objectivity   h) skiing   i) intellect

    j) sexual liberation

Write the corresponding letters below in inverse alphabetical order except for vowels, which also take precedence:

(9) Consider the phrase: 'Science is the way to a white-hot technological future.' Is this true?

If not, what change would you make in the wording?

....

....

SECTION B: Infallibalistic Omnipotentiality Rating

(1) Select the correct form of expression:

a) Food is safe and harmless.

b) Foods may be dangerous in certain respects.

c) In all cases of severe and agonising cancer, there has been evidence that the sufferer had previously been eating food. Need one say more than this?

Place a tick in the appropriate box thus:

   a) ☐   b) ☐   c) ☐

(2) If an 'error' is pointed out in your published work, what was it doing there?

It was missed by a secretary in proof-reading.

It was an inadvertent slip.

It was included in order to test the powers of observation of the reader.

Draw a box around the corresponding cross:

   a) ✗   b) ✗   c) ✗

(3) You are asked for a figure that you do not know off-hand. What is the correct reply to the enquirer?

a) 'I'll look it up if you hang on a moment.'

b) 'Opinions vary as to the definitive amount.'

c) 'Someone of your calibre should have such knowledge at your finger-tips without needing my help to get you out of difficulties.'

d) 'Off-hand I confess I do not know.'

Mark the correct choice with a dodecahedron.

(4) What is the correct terminological definitisation of the following form of statement:

I knew a man who wore a bowler hat in spite of my advice: he died of a stroke at the age of 42.

a) Liberalised objectivity.

b) A rib-tickling leg-pull.

c) A Margument.

d) Illogical absurdity.

e) Juxtapositionary Implicationism.

(5) What is the film from which the Expert's Theme Tune is taken?

2001, A Space Odyssey ☐

Snow White and the Seven Dwarfs ☐

Dr Strangelove ☐

The Andromeda Strain ☐

The Graduate ☐

SECTION C: Supracharacteristic Sociotropism Test

(1) How is your telephone answered at work?

a) 'Hello, technical wing.'

b) 'Good afternoon, this is the director's switchboard.'

c) 'Yeah?'

d) 'This is the Expert's personal private extension, if you wish to leave a message I can put it on the scrambler.'

Underline your choice with a simple sine curve $\lambda \triangleq 0{\cdot}9$ cm.

(2) Select the normal course of events that your overcoat undergoes after you return home:

a) You hang it in the cupboard under the stairs.

b) It is draped over the kitchen bench.

c) Your spouse deferentially takes it from you and puts it immediately on a hanger upstairs.

d) You do not wear one as your chauffeur-driven car is air conditioned. So is the house. *Turn page before continuing*

(3) What do you do if a tradesman comes to the house? Mark your choice:

a) Ignore him and summon somebody else to deal with the matter.

b) Offer him a cigarette/chat/cup of coffee.

c) Deal with him man-to-man and settle it quickly.

d) You would not know as your study is soundproofed and anyway the maid or butler always deals with such things.

(4) Do you smile at people in the street?

a) no　　　b) occasionally　　　c) frequently　　　d) always

Cross out the three that do not apply.

(5) For whom would you most likely be mistaken at a distance?

a) the Duke of Edinburgh ☐

b) the President of the United States ☐

c) the Messiah ☐

SECTION D: Endofacultatively Reorientational Normoprofessionalism

(1) When did you last buy a drink for a newspaper reporter? Subtract number of months that have elapsed from 20 (maximum) and note it down:

(2) Who would you normally invite to a party? Encircle them:

a) the local churchman

b) a film director

c) an old school chum in accountancy

d) a TV production assistant

e) a cub reporter from the local newspaper

f) Gran

g) any Experts younger than yourself

(3) Which is the most correct form of abbreviation for the expression at the top of this section?

a) Chatting up the opposition

b) *Ere*N

c) Endo-waddyacall-it

(4) Is the liberal use of back-slapping humour and general friendliness:

      a) good   b) helpful   c) dangerous

Invert the letter corresponding to your choice and place it in the central hexagon below:

FINAL QUESTION: SECTION E

(1) The rubric said, 'Read all questions carefully before commencing paper.' Have you done this, or have you been completing all the answers already?

      a) Having read the rubric, I am still reading through the Paper as instructed therein.

      b) I have already marked my answers to the questions.

IMPORTANT: Before proceeding first check answers to this question.

ANSWERS

SECTION A:

(1) yes—10   no—2   occasionally—0   frequently—7

(2) score number of marks in box

(3) excellent—10   fair—2   poor—0

(4) score number of marks in box

(5) a—0   b—1   c—10   d—8   e—1

(6) a—0   b—1   c—10   d—9   e—1   f—0

(7) a—2   b—0   c—1   d—10   e—0   f—10

(8) a—4   b—0   c—4   d—0   e—0   f—4   g—0   h—4   i—0   j—4

(9) False—5 points. Correction is: delete science, insert Nonscience—5 points.

NOTE TOTAL:

Quasi-notional Fashionistic Normativity Rating   *qu*FN =   %

SECTION B:

(1) a—0   b—2   c—20

(2) a—15   b—0   c—20

(3) a—2   b—18   c—20   d—0

(4) a—8   b—1   c—17   d—divide your total for this section by 2·4   e—20

(5) a—9   b—20 ('I.O., I.O., as off to work I go.')   c—0   d—1   e—5

NOTE TOTAL: Infallibalistic Omnipotentiality Score   IO =   %

SECTION C:

(1) a—5   b—17   c—0   d—20

(2) a—5   b—0   c—18   d—20

(3) a—5   b—subtract 15 from your total in this section   c—2   d—20

(4) a—15   b—20   c—2   d—0 (you are liable to be arrested and should therefore try to stop this habit)

(5) a—12   b—2   c—20

NOTE TOTAL: Supracharacteristic Sociotropism   SS =   %

SECTION D:

(1) Score number of marks in box

(2) a—0 (in Britain), 15 (in United States), 20 (Europe)   b—15   c—5   d—10   e—20   f—0   g—0

(3) a—6   b—10   c—0 (and you are lucky to get away with that)

(4) a—0   b—1   c—20

NOTE TOTAL:
Endofacultatively Reorientational Normoprofessionalism   EreN =   %

SECTION E:

(1) a—Do not bother to count score, or even to sit paper. You are obviously good Expert material since your brain instinctively followed the instructions to the letter. Take an honorary total of 99% and insert it at the end of the paper as your *Ex* (for Expert) rating.
b—Tch, tch.

TO DETERMINE YOUR EXPERT POTENTIAL

Note your scores, for ease of reference, in the table:

qu FN	
I.O.	
S.S.	
Ere N	
$n^\S$	

Note: $n^\S$ = number of *sections* (above) answered.

Substitute figures obtained in Expertistical Formula:

$$Ex = qu\text{FN} + 2(\text{I.O.} + \text{S.S.})/8 + 2\ Ere\text{N}$$

Your *Ex* (for Expert) rating is now to be marked in the frame of reference:

$$Ex = \qquad \%$$

INTERPRETATION OF RESULTS:

A score of 49·99 or less: you are clearly unfitted to become an Expert and must resign yourself to a more menial—but 'equally valuable'—life as a member of the Lay Public. However, do not worry; we'll take care of everything for you.

A score of 50·00 or more: welcome to the clan! You have a good Expert Potential and clearly have a promising career in front of you. Now read this book again and *memorise it from cover to cover*—it will stand you in good stead.

Index